Natural Gas Production Engineering

Natural Gas Production Engineering

Editor

Alaknanda Sathe

Natural Gas Production Engineering

Edited by **Alaknanda Sathe**

Printed in 2017

ISBN: 978-1-68117-420-4

Library of Congress Control Number: 2015936538

© 2016 by
SCITUS Academics LLC,
616, Corporate Way, Suite 2, 4766,
Valley Cottage, NY 10989

www.scitusacademics.com

This book contains information obtained from highly regarded resources. Copyright for individual articles remains with the authors as indicated. All chapters are distributed under the terms of the Creative Commons Attribution License, which permits unrestricted use, distribution, and reproduction in any medium, provided the original author and source are credited.

Notice

Reasonable efforts have been made to publish reliable data and views articulated in the chapters are those of the individual contributors, and not necessarily those of the editors or publishers. Editors or publishers are not responsible for the accuracy of the information in the published chapters or consequences of their use. The publisher believes no responsibility for any damage or grievance to the persons or property arising out of the use of any materials, instructions, methods or thoughts in the book. The editors and the publisher have attempted to trace the copyright holders of all material reproduced in this publication and apologize to copyright holders if permission has not been obtained. If any copyright holder has not been acknowledged, please write to us so we may rectify.

Contents

Preface vii

Chapter 1   Natural Gas Production from Methane Hydrate Deposits Using CO_2 Clathrate Sequestration: State-of-the-Art Review and New Technical Approaches 1
Annick Nago and Antonio Nieto

Chapter 2   Effect of Bed Deformation on Natural Gas Production from Hydrates 15
Mohamed Iqbal Pallipurath

Chapter 3   Thermochemical Equilibrium Model of Synthetic Natural Gas Production from Coal Gasification Using Aspen Plus 39
Rolando Barrera, Carlos Salazar, and Juan F. Pérez

Chapter 4   Numerical Simulation of an Industrial Absorber for Dehydration of Natural Gas Using Triethylene Glycol 89
Kenneth Kekpugile Dagde and Jackson Gunorubon Akpa

Chapter 5   A Field Study on Simulation of CO_2 Injection and ECBM Production and Prediction of CO_2 Storage Capacity in Unmineable Coal Seam 115
Qin He, Shahab D. Mohaghegh, and Vida Gholami

Chapter 6   Advances in Pressure Swing Adsorption for Gas Separation 135
Carlos A. Grande

Chapter 7   The Role of Natural Fractures in Shale Gas Production 171
Ian Walton and John McLennan

Citations 209
Index 213

Preface

This book presents the quintessential guide for gas engineers, emphasizing the practical aspects of natural gas production and it is primarily written for practicing production engineers. I have made every effort to understand the typical needs of an engineer working in the gas field, and have tried to address them. The book is written as a self-help book. After each concept is illustrated, a numerical example is shown to emphasize the concept, and work problems are provided for a better understanding. Realizing that field units are used in the United States, whereas, rest of the world uses SI units, all the numbers as well as equations used in the book are shown using both units.

Editor

Natural Gas Production from Methane Hydrate Deposits Using CO$_2$ Clathrate Sequestration: State-of-the-Art Review and New Technical Approaches

Annick Nago and Antonio Nieto

The John and Willie Leone Family Department of Energy and Mineral Engineering, The Pennsylvania State University, University Park, PA 16802, USA

ABSTRACT

This paper focuses on reviewing the currently available solutions for natural gas production from methane hydrate deposits using CO$_2$ sequestration. Methane hydrates are ice-like materials, which form at low temperature and high pressure and are located in permafrost areas and oceanic environments. They represent a huge hydrocarbon resource, which could supply the entire world for centuries. Fossil-fuel-based energy is still a major source of carbon dioxide emissions which contribute greatly to the issue of global warming and climate change. Geological sequestration of carbon dioxide appears as the safest and most stable way to reduce such emissions for it involves the trapping of CO_2 into hydrocarbon reservoirs and aquifers. Indeed, CO_2 can also be sequestered as hydrates while helping dissociate the in situ methane hydrates. The studies presented here investigate the molecular exchange between CO_2 and CH_4 that occurs when methane hydrates are exposed to CO_2, thus generating the release of natural gas and the trapping of carbon dioxide as gas clathrate. These projects include laboratory studies on the synthesis, thermodynamics, phase equilibrium, kinetics, cage occupancy, and the methane recovery potential of the mixed CO_2–CH_4 hydrate. An experimental and numerical evaluation of the effect of porous media on the gas exchange is described. Finally, a few field studies on the potential of this new gas hydrate recovery technique are presented.

INTRODUCTION

Since their initial discovery by Sir Davy Humphrey in 1810, natural gas hydrates have graduated from a laboratory oddity to a hydrocarbon production nuisance as seen forming inside the chamber bell used to cap the spill in the deep water horizon oil well, and so forth, before being considered as a potential energy resource for the future. For many decades, countries such as the USA, Canada, Japan, India, and China have funded major research projects to get a better understanding and knowledge of natural gas hydrates [1]. Resource assessment studies have demonstrated the huge potential of gas hydrate accumulations as a future energy resource [2].

World energy demand is steadily rising due to global population and economic growth. World energy consumption is expected to increase from 472 quadrillion Btu to 678 quadrillion Btu in 2030, that a total increase of 44% from 2006 to 2030 [3]. China and India are currently the fastest growing non-OECD economies, and their combined energy consumption is expected to represent 28% of the world energy consumption in 2030 [4]. Despite recent progress in obtaining energy from nonfossil fuels, nearly 80% of the world energy supply will still be generated from oil, natural gas, and coal. The combustion of these fuels is a major source of carbon dioxide emissions. Unfortunately, a perceived change in the global climate has been attributed to the increasing concentration of Green House Gases such as CO_2 in the atmosphere. Geological sequestration of CO_2 is a potential solution to this problem. Typical geological sequestration consists in capturing and storing the gas in a geological setting such as active and depleted oil/gas reservoir, deep brine formations, deep coal seams, and coal-bed methane formation [5]. Sequestration of CO_2 in marine and arctic hydrates is considered as an advanced geologic sequestration concept, which needs further investigation [6].

Gas hydrates are found in nature, in permafrost and marine environments. They contain mixtures of gases such as methane and ethane, with carbon dioxide and hydrogen sulfide as trace. Methane is the predominant component of natural gas hydrates, which is the reason they are simply called methane hydrates. Gas hydrates form under specific conditions: (1) the right combination of pressure and temperature (high pressure and low temperature), (2) the presence of hydrate-forming gas in sufficient amounts, and (3) the presence of water. CO_2 and CH_4 hydrates are of interest with CO_2 being a preferential hydrate guest former when compared to CH_4. In addition, CO_2 hydrates are more stable than CH_4 hydrates, and the exposition CH_4 hydrates to carbon dioxide has resulted in the release of methane, while carbon dioxide remained trapped. Thus, the use of carbon dioxide to recover natural gas from hydrate deposits has gained more and more relevance in the industry.

Other techniques are being explored in the area of production from hydrate deposits. However, the resource is still not commercially viable due to technical, environmental, and economic issues. Any further investigation of the mixed CO_2–CH_4 gas hydrate properties could lead to major breakthroughs in the fields of unconventional resource production and carbon sequestration.

WHAT ARE METHANE HYDRATES?

Natural gas hydrates, commonly called methane hydrates, are crystalline compounds, which are constituted of gas and water molecules. The water molecules or host molecules form a hydrogen-bonded lattice, in which gas molecules or guest molecules are entrapped. The presence of guest molecules stabilizes the lattice due to the sum of the attractive or repulsive forces between molecules known as the Van der Waals forces. There is no bonding between the host molecules and the guest molecules, that is, the gas molecules are free to rotate inside the lattice [2, 7–9]. Gas hydrate formation and dissociation are described by the following equations:

$G + N_H H_2O \rightarrow G.N_H H_2O$ and $G.N_H H_2O \rightarrow G + N_H H_2O$, where N_H is the hydration number and G is the guest molecule Gas hydrate formation is an exothermic process while gas hydrate dissociation is endothermic.

Gas hydrates come under three distinguishable structures: type I, type II, and type H. All structures involve a network of interconnected cages. Structure I (sI) hydrates display unit cells that are constituted of 46 water molecules organized into 2 small cavities and 6 large cavities. The small cavities are dodecahedral cages with 12 pentagonal faces. They are usually denoted as 5^{12} cages. The large cavities are 14-sided polyhedra (tetrakaidecahedron), which are usually denoted as $5^{12}6^2$. The unit cells of Type II hydrates (sII) contain 136 water molecules. They are organized into 16 small cavities and 8 large cavities. The small cavities are of the same kind as the small cavities in sI hydrates. However, the large cavities are hexacaidecahedra ($5^{12}6^4$) with 12 pentagonal faces and 4 hexagonal faces [9]. In 1987, a new hydrate structure was discovered and called structure H (sH). This structure contains 34 water molecules in its unit cell, forming a hexagonal lattice. Type H hydrates display three types of cavities: three 5^{12} cages, two $4^35^66^3$ cages, and one large $5^{12}6^8$ [9, 10].

Because of the size difference between the cages, the three types of hydrates tend to trap different kinds of molecules. Type I hydrates are usually formed with smaller molecules such as ethane and hydrogen sulfide. Type II clathrates are formed by larger molecules such as propane and isobutane. Type H hydrates require the presence of a small molecule such as methane and a type H gas former like 2-methylbutane and cycloheptane to be created. They are less common in nature than the other types of gas hydrates [9, 10]. Figure 1 illustrates the different sorts of hydrate structures and some of their gas-forming molecules. These structures have been observed with X-ray diffraction.

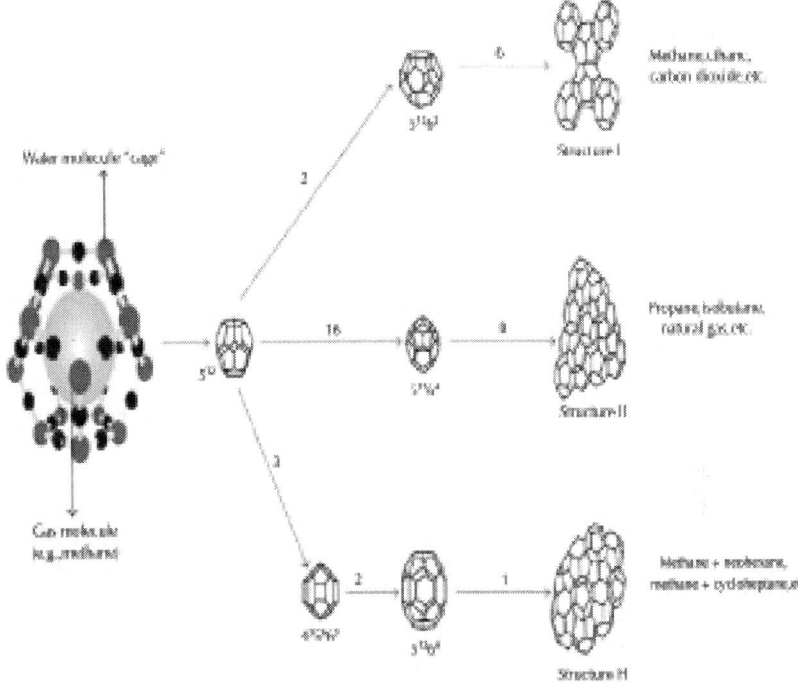

Figure 1: Different types of clathrate hydrates [9].

Methane and carbon dioxide both form type I hydrates. The comparison of their hydrate phase equilibrium conditions suggests the occurrence of a transition zone between both hydrate equilibrium curves where CO_2 hydrates can exist while CH_4 hydrates dissociate into methane gas and water. The hydrate phase diagrams of both compounds are presented in Figure 2. In addition, the heat of formation of carbon dioxide hydrate (−57.98 kJ/mole) is greater than the heat of dissociation of methane hydrate (54.49 kJ/mole) The heat released from the formation of carbon dioxide hydrate in the presence of methane hydrate should be sufficient to dissociate the methane hydrate and recover methane gas [11]. Thirdly, it has been experimentally proven that carbon dioxide is preferentially trapped over methane in the hydrate phase [12]. These observations fuel the growing interest in the use of carbon dioxide for natural gas recovery from gas hydrate deposits.

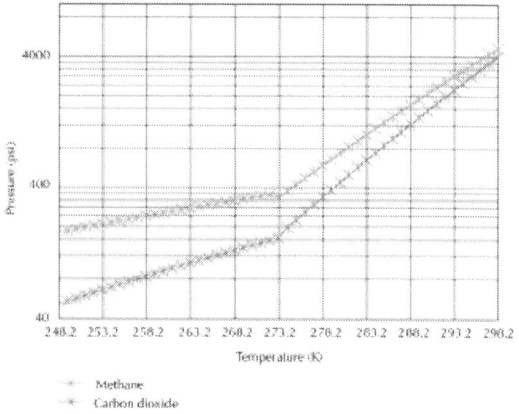

Figure 2: CH_4 and CO_2 hydrate phase diagrams [2].

Gas hydrates can be naturally found in permafrost areas and subsea environments. The temperature and pressure gradients which are at play underneath the Earth help define specific hydrate occurring zones, when associated to the thermodynamic hydrate equilibrium conditions. These zones are called hydrate stability zones [8]. Figure 3 displays the hydrate stability zones in permafrost and marine environments.

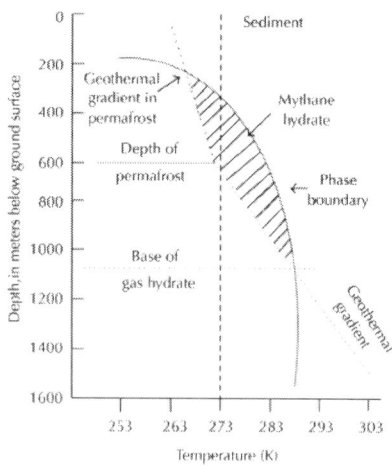

(a)

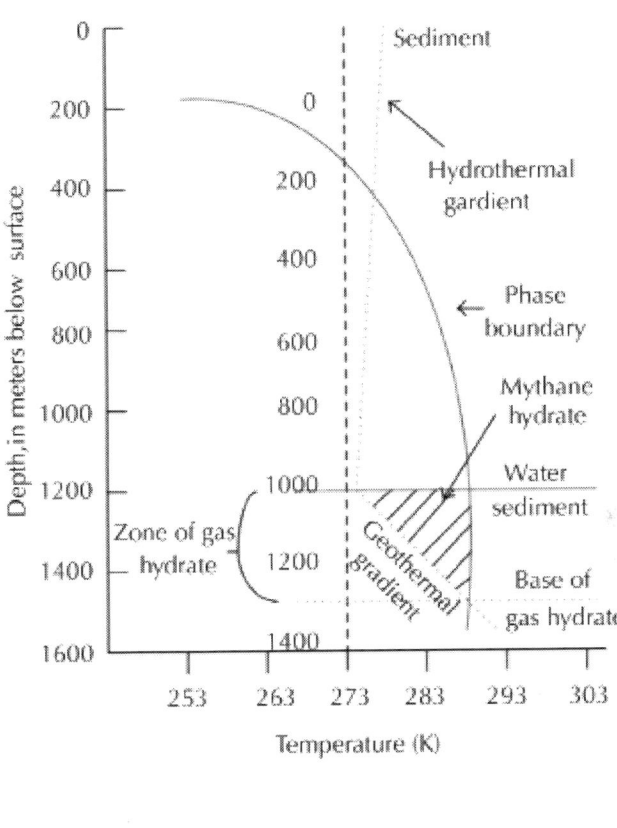

(b)

Figure 3: Hydrate stability zones in permafrost and marine environments [2].

Assessment methods for gas hydrates include seismic studies (bottom simulating reflectors), pore water salinity measurements, well-logging, and direct observations from core samples [13]. So far, 89 hydrate locations have been discovered all over the world [14]. These locations are presented in Figure 4.

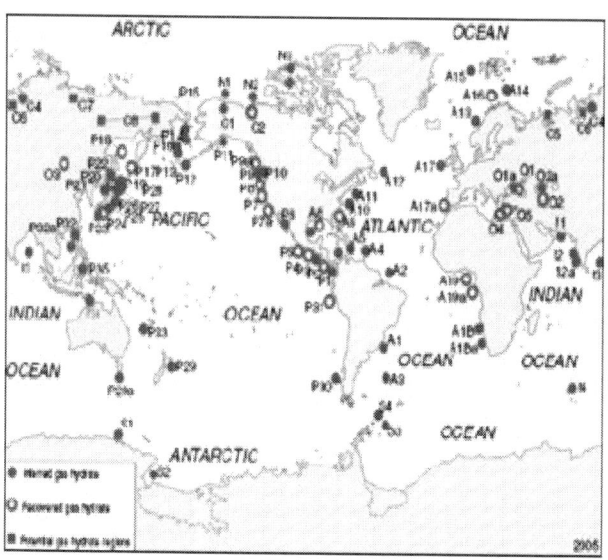

Figure 4: World gas hydrate locations [13].

CURRENT RESEARCH STATUS

Three main production methods have so far been explored for the recovery of natural gas from hydrate deposits: depressurization, thermal stimulation, and inhibitor injection [8, 15, 16]. These methods aim at thermodynamically destabilizing the reservoir environment to provoke the release of the entrapped gas [17, 18]. They have been investigated experimentally, numerically, and in the field. However, they have not yet been used for commercial production of natural gas hydrates due to remaining technical and economic issues. A fourth method was introduced a few years ago and is based on the concept of hydrate guest molecule exchange between methane and carbon dioxide in the hydrate phase.

In 1996, Ohgaki et al. [12] examined the possible interactions between these two hydrates by injecting carbon dioxide (gas) into an aqueous-gas hydrate system containing methane. CO_2 displays a higher chemical affinity than CH_4 in the hydrate structure since it has a higher heat of formation and equilibrium temperature; that is, at 1000 psi, the equilibrium temperature of CH_4 hydrate is approximately

283.15 K while the equilibrium temperature of CO_2 hydrate is around 286.15 K. Ohgaki et al.'s experiments resulted in the synthesis of a mixed CO_2–CH_4 hydrate. The equilibrium concentrations obtained for CO_2 were greater in the hydrate phase than those of CH_4 and less than the concentrations of CH_4 in the gas phase. Nakano et al. (1998) [19] performed a similar study using carbon dioxide and ethane and obtained comparable results. Smith et al. (2001) [20] inquired the feasibility of exchanging carbon dioxide with methane in geologic accumulations of natural gas hydrates. They numerically investigated the effect of the pore size distribution on the conversion of CH_4 hydrate to CO_2 hydrate. It was demonstrated that the guest molecule exchange between CO_2 and CH_4, in porous media was less thermodynamically favored, as the pore size decreased. They recommended these numerical results be validated by laboratory experiments. Seo et al. (2001) [21] experimentally investigated hydrate phase equilibrium processes for mixtures of CO_2 and CH_4. They determined the existing conditions of quadruple points (H-L_w-L_{CO2}-V) in order to examine the hydrate stability. It was noted that the equilibrium curves of the mixed hydrates lied between those of simple carbon dioxide and methane hydrates. For a given mixture, the concentration of CO_2 in the hydrate phase decreased as the pressure was lowered. In 2003, Lee et al. [22] published the results of their study on the thermodynamics and kinetics of the conversion of CH_4 hydrate to CO_2 hydrate. They analyzed the distribution of guest molecules over different cavities for pure methane hydrates and different mixtures of CO_2–CH_4 hydrates, using solid state NMR methods. It was observed that the cage occupancy ratio of CH_4 in the pure methane hydrate decreased as the concentration of CO_2 in the mixture increased. This was explained by the fact that CO_2 preferentially occupied large $5^{12}6^2$ cages in the mixed hydrate. In terms of kinetics, it was noticed that the conversion of CH_4 hydrate to CO_2 hydrate happened much more quickly than the formations of pure CO_2 and CH_4 hydrates. The amount of CH_4 that could be recovered from the gas hydrate of composition CH_4 6.05H_2O was limited to 64% of the original entrapped gas, even with a CO_2 concentration of 100 mol%. Ota et al. (2004) [23] focused on the gas exchange process using liquid CO_2. They performed laboratory measurements using the Raman spectroscopy and numerical simulations, and they found similar results in terms of feasibility of the molecular gas exchange. Stevens et al. (2008) [24] took the studies on this topic one step further by publishing his work on the gas exchange between CO_2 and CH_4 in hydrates formed within sandstone core samples. He used a MRI to analyze the samples and realized there was formation of CO_2 hydrate at the expense of the initial CH_4 hydrate. Diffusion seemed to be the main driving force behind the conversion from CH_4 hydrate to CO_2 hydrate. A considerable amount of CH_4 was released during the process, which was judged as rapid and efficient. There was no free water present. The permeability of the core was reduced during CH_4 hydrate formation. This reduced permeability was maintained constant during the CH_4–CO_2 exchange, and the permeability levels were considered sufficient for gas transportation. In 2008, Young June et al. [25] made a major discovery while they were inquiring the effect of the injection of a binary mixture of N_2 and CO_2 on methane hydrate recovery. They found out that the injection of a binary mixture of N_2 and CO_2, instead of the traditional pure CO_2, increased the percentage of methane recovered from 64% to 85% for type I gas hydrates. They also looked at the potential influence of structural transition by forming a type II CH_4–C_2H_6 hydrate and injecting CO_2 and a mixture of CO_2 with N_2. It was determined that the hydrate structure changed from type II to type I during the gas injection, thus increasing the gas recovery to more than 90% for CH_4. Besides these major thermodynamically related numerical and laboratory investigations, several studies were conducted to evaluate the potential of this new concept as a field scale production method for methane hydrate deposits. In 2003, Rice [26] proposed a scheme for methane recovery from marine hydrate accumulations. In this scheme, the produced methane would be converted into hydrogen and carbon dioxide; then, the carbon dioxide would be reinjected into the ocean to be converted into CO_2 hydrates and finally the produced hydrogen would be used as fuel. Methane would be recovered from hydrates using depressurization combined with thermal stimulation. No direct molecular gas exchange between CH_4 and CO_2 was inferred in this production scheme. In 2004, McGrail et al. [27] investigated Ohgaki et al.'s method to determine the rate of CO_2 gas penetration in the bulk methane hydrate, using the Raman spectroscopy. They discovered that the rates of CO_2 gas penetration were too low for this method to be useful for gas hydrate production. Then, they performed a preliminary study on a new enhanced gas hydrate recovery concept based on the injection of a microemulsion of CO_2 and water in the methane hydrate core samples. The technique was validated through laboratory experiments and numerical simulation, using a custom model based on STOMP-CO_2. Finally, Castaldi et al. (2006) [28] examined the technical feasibility of applying a down-hole combustion method for gas recovery from hydrate accumulations, while sequestrating CO_2 as hydrates. The gas molecular exchange between CH_4 and CO_2 was not directly mentioned, but they suggested there should be equality between the rates of CO_2 hydrate formation and CH_4 hydrate dissociation, during the process. In 2006, Goel [11] released a review of the status of research projects and issues related to methane hydrate production with carbon dioxide sequestration. It was concluded that although several studies had been performed on the topic, additional experimental data was needed on the topic of CH_4–CO_2 molecular gas exchange in hydrate-bearing sediments. He emphasized the importance of fully knowing the thermodynamics and kinetics of the formation and dissociation of this mixed hydrate and of the conversion process, in porous media. He also pointed out the essence of understanding the equilibrium conditions of the mixed hydrate in sediments as a function of pressure, temperature, mole fraction of CO_2 and CH_4 in the mixture, pore size, porous material, and flow properties.

CONCLUSIONS

This paper is a brief review of the studies that have been performed on the gas molecular exchange between CO_2 and CH_4 within the hydrate phase. As this paper highlights, such studies are even more essential in this day and age, as we need to quickly discover and exploit new sources of energy in a sustainable and energy-efficient manner. An emphasis is put here on experimental, numerical, and field investigations of the gas hydrate recovery process using CO_2, clathrate sequestration. All studies present positive outcomes and further research on the topic is encouraged to make this new recovery technique commercially viable.

REFERENCES

1. Committee to Review the Activities Authorized Under the Methane Hydrate Research and Development Act of 2000,

Charting the Future of Methane Hydrate Research in the United States, The National Academies Press, Washington, DC, USA, 2004.
2. S. E. Dendy and C. A. Koh, Clathrate Hydrates of Natural Gas, CRC Press, Boca Raton, Fla, USA, 2008.
3. Exxon Mobil Corporation, The Outlook for Energy: A View to 2030, 2009.
4. EIA, Energy Information Administration-Official Energy Information Administration website, 2009, http://www.eia.gov/.
5. D. A. Voormeij and G. J. Simandl, "Geological, ocean, and mineral CO_2 sequestration options: a technical review," Geoscience Canada, vol. 31, no. 1, pp. 11–22, 2004.
6. DOE's Office of Fossil Energy and Office of Science, "Carbon sequestration research and development," US Department of Energy, 1999.
7. M. R. Prado, A. Pham, R. E. Ferazzi, K. Edwards, and K. C. Janda, "Gas clathrate hydrates experiment for high school projects and undergraduate laboratories," Journal of Chemical Education, vol. 84, no. 11, pp. 1790–1791, 2007.
8. Y. F. Makogon, Hydrates of Hydrocarbons, Pennwell Publishing Company, Tulsa, Okla, USA, 1997.
9. "Center for gas hydrate research," Herriot-Watt University, Institute of Petroleum Engineering, 2010, http://www.pet.hw.ac.uk/research/hydrate/index.cfm.
10. J. Caroll, Natural Gas Hydrate: A Guide for Engineers, Gulf Professional Publishing, 2003.
11. N. Goel, "In situ methane hydrate dissociation with carbon dioxide sequestration: current knowledge and issues," Journal of Petroleum Science and Engineering, vol. 51, no. 3-4, pp. 169–184, 2006.
12. K. Ohgaki, K. Takano, H. Sangawa, T. Matsubara, and S. Nakano, "Methane exploitation by carbon dioxide from gas hydrates-phase equilibria for CO_2-CH_4 mixed hydrate system," Journal of Chemical Engineering of Japan, vol. 29, no. 3, pp. 478–483, 1996.
13. G. J. Moridis, T. S. Collett, R. Boswell et al., "Toward production from gas hydrates: current status, assessment of resources, and

simulation-based evaluation of technology and potential," SPE Reservoir Evaluation and Engineering, vol. 12, no. 5, pp. 745–771, 2009.
14. K. A. Kvenvolden and B. W. Rogers, "Gaia's breath-global methane exhalations," Marine and Petroleum Geology, vol. 22, no. 4, pp. 579–590, 2005.
15. H. A. Phale, T. Zhu, M. D. White, and B. P. McGrail, "Simulation study on injection of CO_2-microemulsion for methane recovery from gas-hydrate reservoirs," in Proceedings of the 2006 SPE Gas Technology Symposium, pp. 369–380, Society of Petroleum Engineers, Calgary, Canada, May 2006.
16. M. D. Max and M. J. Cruickshank, "Extraction of methane from oceanic hydrate system deposits," inProceedings of the Offshore Technology Conference, Houston, Tex, USA, 1999.
17. C. P. Thomas, "Methane hydrates: major energy source for the future of wishful thinking?" in Proceedings of the 2001 SPE Annual Technical Conference and Exhibition, New Orleans, La, USA, 2001.
18. M. D. White, S. K. Wurstner, and B. P. McGrail, "Numerical studies of methane production from Class 1 gas hydrate accumulations enhanced with carbon dioxide injection," Marine and Petroleum Geology, vol. 28, no. 2, pp. 546–560, 2011.
19. S. Nakano, K. Yamamoto, and K. Ohgaki, "Natural gas exploitation by carbon dioxide from gas hydrate fields-high-pressure phase equilibrium for an ethane hydrate system," Proceedings of the Institution of Mechanical Engineers, Part A, vol. 212, no. 3, pp. 159–163, 1998.
20. D. H. Smith, K. Seshadri, and J. W. Wilder, "Assessing the thermodynamic feasibility of the conversion of methane hydrate into carbon dioxide hydrate in porous media," Journal of Energy and Environmental Research, pp. 101–117, 2001.
21. Y.-T. Seo, H. Lee, and J. H. Yoon, "Hydrate phase equilibria of the carbon dioxide, methane, and water system," Journal of Chemical and Engineering Data, vol. 46, no. 2, pp. 381–384, 2001.
22. H. Lee, Y.-T. Seo, I. L. Moudrakovski, and J. A. Ripmeester, "Recovering methane from solid methane hydrate with carbon dioxide," Angewandte Chemie, vol. 42, no. 41, pp. 5048–5051, 2003.

23. M. Ota, K. Morohashi, Y. Abe, M. Watanabe, R. Lee Smith Jr., and H. Inomata, "Replacement of CH_4 in the hydrate by use of liquid CO_2," Energy Conversion and Management, vol. 46, no. 11-12, pp. 1680–1691, 2005.
24. J. C. Stevens, J. J. Howard, B. A. Baldwin, G. Ersland, J. Husebo, and A. Graue, "Experimental hydrate formation and gas production scenarios based on CO_2 sequestration," in Proceedings of the 6th International Conference on Gas Hydrates (ICGH '08), Vancouver, Canada, 2008.
25. P. Youngjune, et al., "Swapping carbon dioxide for complex gas hydrate structures," in Proceedings of the 6th International Conference on Gas Hydrates (ICGH '08), Vancouver, Canada, 2008.
26. W. Rice, "Proposed system for hydrogen production from methane hydrate with sequestering of carbon dioxide hydrate," Journal of Energy Resources Technology, vol. 125, no. 4, pp. 253–257, 2003.
27. B. P. McGrail, T. Zhu, R. B. Hunter, M. D. White, S. L. Patil, and A. S. Kulkarni, "A new method for enhanced production of gas hydrate with CO_2," in Proceedings of the AAPG Hedberg Conference, Vancouver, Canada, 2004.
28. M. J. Castaldi, Y. Zhou, and T. M. Yegulalp, "Down-hole combustion method for gas production from methane hydrates," Journal of Petroleum Science and Engineering, vol. 56, no. 1–3, pp. 176–185, 2007.

Chapter 2

Effect of Bed Deformation on Natural Gas Production from Hydrates

Mohamed Iqbal Pallipurath

Mechanical Engineering Department, TKM College of Engineering, Kollam, Kerala 691005, India

ABSTRACT

This work is based on modelling studies in an axisymmetric framework. The thermal stimulation of hydrated sediment is taken to occur by a centrally placed heat source. The model includes the hydrate dissociation and its effect on sediment bed deformation and resulting effect on gas production. A finite element package was customized to simulate the gas production from natural gas hydrate by considering the deformation of submarine bed. Three sediment models have been used to simulate gas production. The effect of sediment deformation on gas production by thermal stimulation is studied. Gas production rate is found to increase with an increase in the source temperature. Porosity of the sediment and saturation of the hydrate both have been found to significantly influence the rate of gas production.

INTRODUCTION

Energy demand is on the rise globally but the production rates of major fossil fuels are going down. Several analysts predict a drastic reduction in energy production due to diminishing reserve of fossil fuels. The major result from the global analysis is that world oil production peaked in 2006. Production has started to decline at a rate of several percentages per year. This necessitates a search for commercially viable and clean source of energy capable of meeting future energy demands. Natural gas hydrate (NGH) is one of the possible energy sources to meet these requirements. It is a highly condensed form of natural gas formed by capture of natural gas molecules in a cage of water molecules: each cubic meter of natural gas hydrate yields about 160 cubic meter of gas at STP.

A large amount of natural gas hydrate exists on our planet. Such deposits are found both on land (in the permafrost region), and offshore (in the submarine sediment). Over 230 gas hydrate deposits have been found globally. Gas hydrates have also been located in the coastal regions of India [1]. Needless to say, the vastness of gas hydrates has attracted global attention for its exploration and exploitation for future energy supply. It is predicted that utilization of even 17% to 20% of this resource could meet the energy demands for next 200 years [2].

Methods suggested for the production of natural gas from gas hydrate include depressurization, thermal stimulation, chemical inhibitor injection, and CO_2 sequestration. Among these, depressurization and thermal stimulation have been considered to be the most economical, though other methods are under investigation. The type of method depends on the reservoir characteristics. Due to less energy input for depressurization, this method has been studied more than the thermal stimulation. However, the efficacy of latter method needs to be studied in more detail. In recent times, during December 2001 to March 2002, field tests were conducted at a permafrost region located at Mallik gas hydrate site (Canada) in which both depressurization and thermal stimulation were conducted. Each of these methods has its relative merits and demerits, and for the development of commercial technology it is essential to have a good understanding of each of the proposed methods.

CURRENT SCENARIO

The study of any of the methods for gas production from natural gas hydrates needs not only a proper knowledge of the thermodynamics, kinetics, and heat and mass transfer effects, but also that of sediment response to any change in its morphology during hydrate dissociation. The presence of hydrate has a cementing effect on the sediment structure and hence any loss of hydrate tends to weaken the sediment. Sediment deformation has profound influence on the gas productivity. Moreover, a reduction in the sediment strength may destabilize sediment matrix to an extent that could cause serious damage to the production operation and uncontrolled release of methane to the environment.

The earlier modeling on gas production from gas hydrates considered an undeformed bed as the main objective was to study the effectiveness of a given method in the gas production. Some of these works considered the kinetics of hydrate dissociation (e.g., [3]), and others assumed instantaneous dissociation and hence equilibrium condition during hydrate dissociation (e.g., [4]). Also, the heat transfer effect was not considered in all the studies (e.g., [3] assumed isothermal operation). Analytical solutions were obtained by Selim and Sloan [4], Yousif et al. [3], Tsypkin [5], Ji et al. [6], and many others. However, such solutions were based on assumptions which were found not to be valid in reality. Numerical approach to the problem has been attempted to address the real conditions existing in the reservoir and other operational issues. Gas production from hydrate reservoir by the combination of warm water flooding and depressurization was proposed by Bai and Li [7] which can overcome the deficiency of single production method. Sun et al. [8] developed a one-dimensional model of hydrate depressurization in porous media.

Recently, a research group led by Professor Kimoto [9, 10] from Kyoto University, Japan, reported studies on gas production from gas hydrates considering bed deformation. They proposed a model based on chemothermomechanically coupled analysis which could predict the deformation of sediment; they did not report the gas production under this condition. Gas production from gas hydrate was also studied by some researchers by using some of the reservoir simulators such as CMG-STARS, Hydrate Res Sim, MH-21 HYDRES, STOMP-HYD, and TOUGH-HYDRATE.

Kinetics and thermodynamics of hydrate formation and dissociation dictate the choice of operating conditions and hence the gas production. Pioneering work on methane hydrate kinetics was done by Professor Bishnoi and his research team from University of Calgary (e.g., [11–13]). Other studies on methane hydrate kinetics were performed with different additives such as polymeric inhibitor [14], electrolyte solution [15], and promoter [16] A kinetic rate model was proposed by Kim et al. [13] for methane hydrate decomposition. Kinetics of hydrate formation of gases other than CH_4, such as C_2H_6, and natural Gas, was studied by Kaschiev and Firoozabadi [17].

The gas hydrate thermodynamics dictates the pressure-temperature relationship to predict the zone of hydrate stability. Also, the heat of dissociation of hydrate is obtained from such studies. Selim and Sloan [4] reported a thermodynamic relationship based on Antoine equation and also an equation to determine the heat of hydrate dissociation. Lu and Sultan (2008) summarized the studies on hydrate stability and proposed a correlation to relate pressure and temperature at hydrate equilibrium which gave a good match with the previous experimental data.

The effects of some polymers and surfactants on methane hydrate formation were investigated by Karaaslan and Parlaktuna (2004) in a high-pressure system.

Results of field test on gas production by thermal stimulation of hydrated sediments performed at Mallik 5L-38 gas hydrate production research well were reported by Hancock et al. [18]. No other field data on thermal stimulation is available.

A review of the reported studies in the literature shows that the following issues need to be investigated in more detail.

- The type of soil mechanical model that can best predict the production behaviour during thermal stimulation has not previously been investigated. The previous work available in the literature on the effect of soil deformation indicates that soil model may have a significant bearing on the gas production.
- Till now, experimental data are available only from one field test on thermal stimulation in permafrost region. It is necessary to conduct experiments to study this method under submarine conditions.

Description of the Model Used

Numerical modelling can be an effective tool that enables understanding mechanisms leading to wellbore instability in oceanic hydrate bearing sediment. To assess deformations caused by hydrate dissociation and the effect of these deformations on the gas generation, numerical techniques are essential. Thermal dissociation of hydrated sediment by a pumped hot fluid is modeled. A radial heat flow from the hot pipe is assumed. The coordinate system is cylindrical. Four components (soil, hydrate, gas (methane), and water) and three phases (hydrate, gas, and aqueous-phase) are considered in the simulator. The intrinsic kinetics of hydrate formation or dissociation is considered using the Kim-Bishnoi model. Mass transport and heat transfer involved in formation or dissociation of hydrates are included in the governing equations. The arrangement of heat source is shown in Figures 1 and 2.

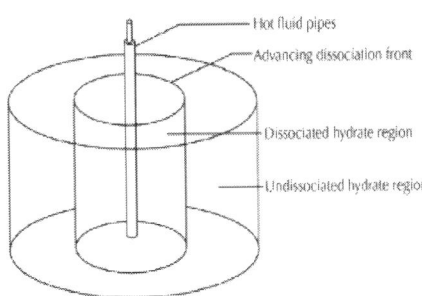

Figure 1: Schematic of a hydrate reservoir heated with pipes carrying hot fluid.

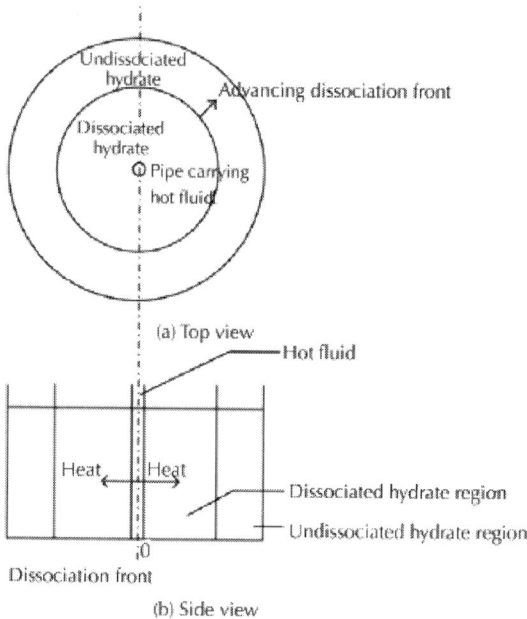

Figure 2: Schematic representation of hydrated sediment to study gas hydrate dissociation by thermal stimulation.

Factors to be evaluated are as follows.
- Changing the stresses and pore pressures.
- Impact of the selected constitutive model for pore pressures.
- Effect of heating the formation on thermodynamic stability of the hydrates.
- Deformation of the sediment as a result of dissociation of hydrate.
- Effect of this deformation on the gas production rates.
- Effect of different saturations of hydrate on gas production.
- Effect of source temperature on gas production.
- Effect of porosity on gas production.

Mechanisms to be considered are as follows.
- Kinetic rate, heat and mass transfer equilibrium, and fluid flow relations for gas hydrate dissociation/reformation with the change in pressure and temperature.

- Resultant changes in the mechanical and petrophysical properties of the sediments.
- Representative constitutive equation and yield criterion for the mechanical behavior of HBS of various hydrate concentrations.
- Different soil models and their responses to deformation.

The generation rates of gas and water are dictated by stoichiometry of the hydrate. The transition between gas, water, and hydrates can be represented as a chemical reaction

$$g\left(\frac{G}{A}\right) + n \cdot w(A) \Longleftrightarrow h(H) \qquad (1)$$

Where g is the gas component, existing as free gas (G) or dissolved in water (A), is the hydrate component present only in the hydrate phase (H), and n is the hydration number.

Assumptions

- Gas hydrate bearing zone at a total depth of 3000 to 3300 meters (~700 m below sea floor and 2400 m below sea level) is considered the default range of study (Yun et al., 2010).
- We neglect the adsorption of any component by the rock phase; that is, the rock phase is inactive in mass transfer.
- Momentum (fluid flow) and heat transfer are axisymmetric.
- The water phase is incompressible.
- The gas follows the Peng Robinson equation of state.
- The porous medium (rock) is nondeformed.
- Gas can occur only in gaseous and hydrate states since CH_4 solubility under conditions of model is negligible.
- Water can occur only in liquid and hydrate states; that is, ice and water vapour formation are neglected since the sediment conditions preclude its formation.

Equivalent thermal conductivity of hydrated sediment is given by the equation

$$\lambda = (1-\varphi)\lambda_s + \sum_{j=h,g,w} \phi S_j \lambda_j. \qquad (2)$$

For a porosity of 0.47, let us consider two extreme cases of hydrate saturation, namely, 0.8 and 0.1. For hydrate saturation of 0.8 and water and gas saturations of 0.1 each, the equivalent thermal conductivity is 3.968 W m^{-1} K^{-1} whereas for a hydrate saturation of 0.1 and water and gas saturations of 0.45 each, the equivalent thermal conductivity is 3.868 W m^{-1} K^{-1} which is a difference of just 2.53%. This shows that an assumption of invariant equivalent thermal conductivity is valid.

Flow through Porous Media Applied to Hydrate Bearing Sediment

A porous medium is modelled in Abaqus/Standard by a conventional approach that considers the medium as a multiphase material and adopts an effective stress principle to describe its behaviour. The porous medium modelling provided considers the presence of two fluids in the medium. One is the "wetting liquid," which is assumed to be relatively (but not entirely) incompressible. Often the other is a gas, which is relatively compressible. An example of such a system is marine hydrated sediment containing sea water and gas. When the medium is partially saturated, both fluids exist at a point; when it is fully saturated, the voids are completely filled with the wetting liquid.

The porous medium is modelled by attaching the finite element mesh to the solid phase; fluid can flow through this mesh.

Coupled Flow and Heat Transfer through Porous Media

Optionally, heat transfer due to conduction in the soil skeleton and pore fluid, as well as convection in the pore fluid, can also be modeled. This capability represents an enhancement to the basic pore fluid flow capabilities discussed in the earlier paragraphs and requires the use of coupled temperature-pore pressure elements that have temperature as an additional degree of freedom in addition to the pore pressure and the displacement components. When you use the coupled temperature-pore pressure elements, Abaqus solves the heat transfer equation in addition to and in a fully coupled manner with the continuity equation and the mechanical equilibrium equations. Only linear brick, first-order axisymmetric, and second-order modified tetrahedrons are available for modeling coupled heat transfer with pore fluid flow and mechanical deformation. Coupled temperature-pore pressure elements are not supported in Abaqus/CAE.

Total and Excess Pore Fluid Pressure

The coupled pore fluid diffusion/stress analysis capability can provide solutions either in terms of total or "excess" pore fluid pressure. The excess pore fluid pressure at a point is the pore fluid pressure in excess of the hydrostatic pressure required to support the weight of pore fluid above the elevation of the material point. The difference between total and excess pore pressure is relevant only for cases in which gravitational loading is important. Total pore pressure solutions are provided when the gravity distributed load is used to define the gravity load on the model. Excess pore pressure solutions are provided in all other cases, for example, when gravity loading is defined with body force distributed loads.

Transient Analysis

In this transient coupled pore pressure/effective stress analysis, the backward difference operator is used to integrate the continuity equation and the heat transfer equation: this operator provides unconditional stability so that the only concern with respect to time integration is accuracy.

For fully saturated flow analyses in which heat transfer is also modelled, the contributions to the model's stiffness matrix arising from convective heat transfer due to pore fluid flow are unsymmetric.

Partially Saturated Flow

In gas hydrate sediment analysis, we shall be dealing with partially saturated flow. In partially saturated flow cases, the corresponding guideline for the minimum time increment is

$$\Delta t > \frac{\gamma_w n^0 (1 + \beta v_w)}{6 k_s k} \frac{ds}{du_w} (\Delta \ell)^2 \quad (3)$$

where s is the saturation, K_s is the permeability-saturation relationship, ds/du_w is the rate of change of saturation with respect to pore pressure, n^0 is the initial porosity of the material, Dt is the time increment, g_w is the specific weight of the wetting liquid, k is the permeability of the soil, v_w is the magnitude of the velocity of the pore fluid, β is the velocity coefficient in Forchheimer's flow law (β=0) in the case of Darcy flow, and $\Delta \ell$ is a typical element dimension.

Automatic Incrementation

Since automatic time incrementation is left to Abaqus, three tolerance parameters are chosen. The accuracy of the time integration of the flow continuity equations is governed by the maximum wetting liquid pore pressure change, Δu_w^{max}, allowed in an increment. Abaqus/Standard restricts the time increments to ensure that this value is not exceeded at any node (except nodes with boundary conditions) during any increment in the analysis.

Since heat transfer is modelled, the accuracy of time integration is also governed by the maximum temperature change, $\Delta \theta_{max}$, allowed in an increment. Abaqus/Standard restricts the time increments to ensure that this value is not exceeded at any node (except nodes with boundary conditions) during any increment of the analysis.

Mechanical Constitutive Models

The constitutive library provided in Abaqus contains a range of linear and nonlinear material models for all of these categories of materials. In general, the library has been developed to provide those models that are most usually required for practical applications. There are several distinct models in the library, and for the more commonly encountered materials, several ways of modeling the material are provided, each suitable to a particular type of analysis application. But the library is far from comprehensive: the range of physical material behavior is far too broad for this ever to be possible. If there is no model in the library that is useful for a particular case, Abaqus/Standard contains a user subroutine UMAT. In these routines the user can code a material model (or call other routines that perform that task). This "user subroutine" capability proved to be a powerful resource for the modelling of gas hydrate bearing sediment.

From a numerical viewpoint, the implementation of a constitutive model involves the integration of the state of the material at an integration point over a time increment during a nonlinear analysis (the implementation of constitutive models in Abaqus assumes that the material behavior is entirely defined by local effects, so each spatial integration point can be treated independently). Since Abaqus/Standard is most commonly used with implicit time integration, the implementation must also provide an accurate "material stiffness matrix" for use in forming the Jacobian of the nonlinear equilibrium equations.

FEM Formulation of Gas Production from Sediment Bed

The Abaqus package was customized with user subroutines and modified input files to solve the model equations. Some typical results obtained for submarine sediment are given. Figure 3 is a comparison of the three developed soil models for the base case using simulation of cumulative gas production for comparison. The three models of soil dynamical system give comparable results as shown in Figure 3.

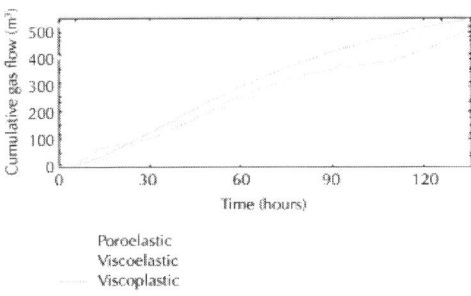

Figure 3: Three soil models compared in cumulative gas production rate.

In Figures 4 and 5, the temperature profiles for the viscoplastic and viscoelastic soil models are compared, and the figures indicate that, by around 60 hours, the temperature profile steadies and is unvarying with time thereafter. It may be inferred that, for a given radius of reservoir, there exists a constant period of time after which the temperature profile can be said to have achieved steady state.

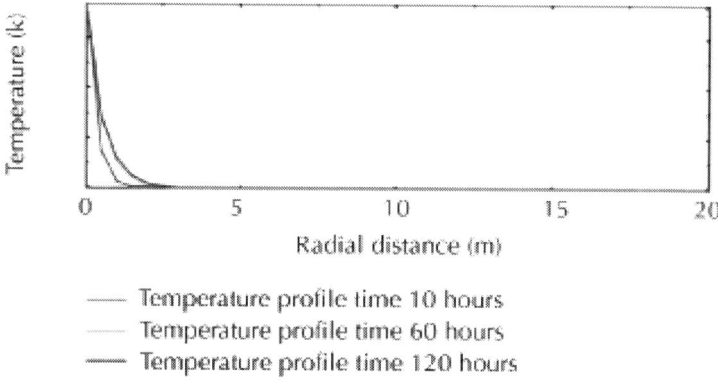

Figure 4: Temperature profile in sediment at various times using viscoplastic model.

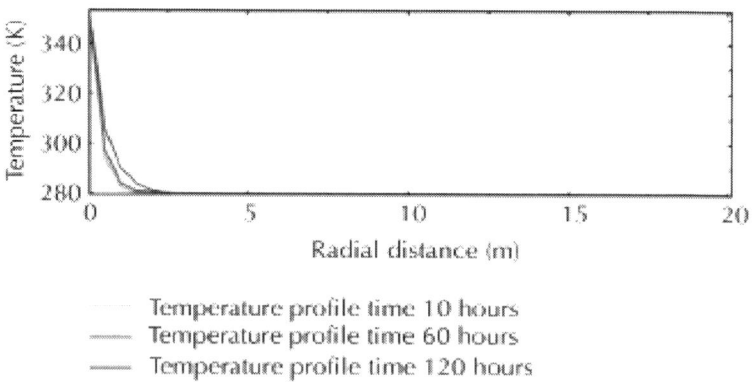

Figure 5: Temperature profile in sediment at various times using viscoelastic model.

It appears from the curves in Figures 4 and 5 that the steady state temperature distribution is not dependent on the soil model, provided the thermal properties of the sediment are more or less constant, as they are in these two soil models. But, as can be seen, the rates at which the steady states are achieved can vary marginally with the soil models used.

The effects of deformation on gas production are shown in Figures 6 and 7. Figure 6 shows cumulative gas production without deformation. When the plots are compared side by side, it appears that the amount of gas produced is reduced by more than 50% when we allow for deformation of the sediment. This could be explained by the reduced permeability of the sediment after dissociation and subsequent subsidence.

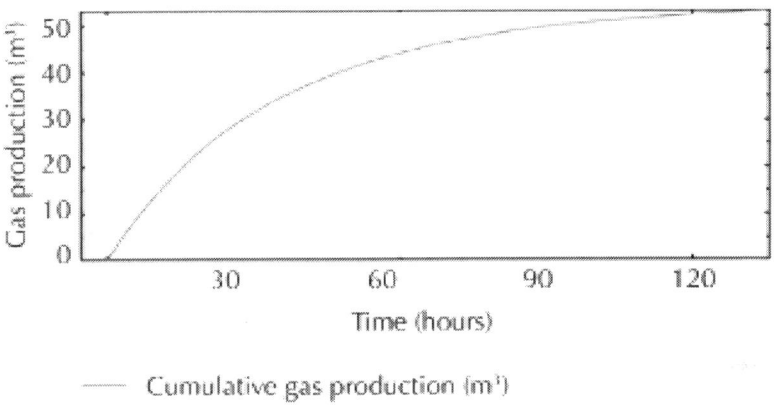

—— Cumulative gas production (m³)

Figure 6: Cumulative gas production without deformation.

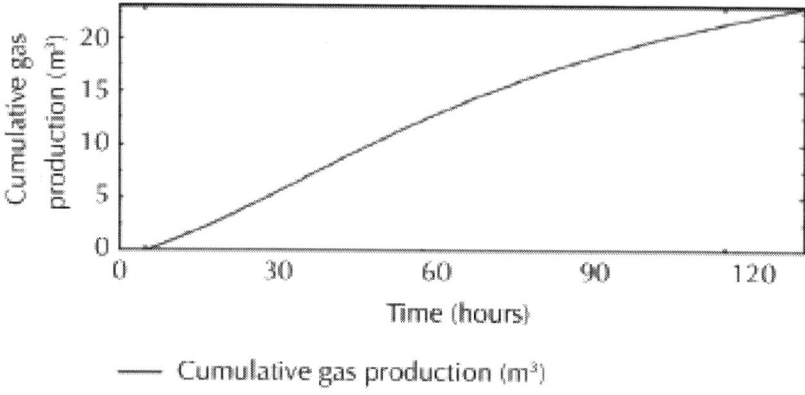

Figure 7: Cumulative gas production with deformation (viscoelastic soil model).

Figure 7 shows cumulative gas production with deformation of sediment bed using the viscoelastic soil model. The gaps between the sediment grains are reduced after subsidence, which could account for the drop in gas/water permeability, which in turn reduces the gas production by more than half.

Parametric Study

A few representative results are shown of parametric studies conducted to gauge the impact of porosity and hydrate saturation upon gas production rate. The effect of sediment porosity upon gas production is graphically presented in Figure 8. Higher porosity seems to aid the production rate due to greater mobility for gas in higher porosity sediment (with higher permeability).

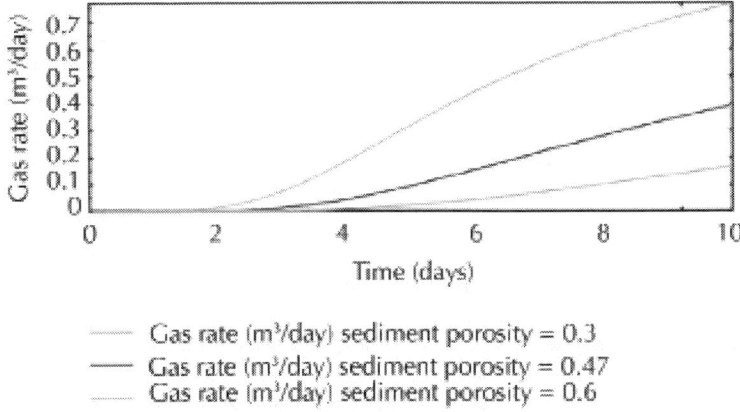

- - - Gas rate (m³/day) sediment porosity = 0.3
—— Gas rate (m³/day) sediment porosity = 0.47
······ Gas rate (m³/day) sediment porosity = 0.6

Figure 8: Effect of initial bed porosity on gas production rate.

Gas production rates at different initial hydrate saturations are shown in Figure 9. The production rates are dependent on initial hydrate saturation in a way that is expected, with higher saturations yielding greater amount of gas produced per day.

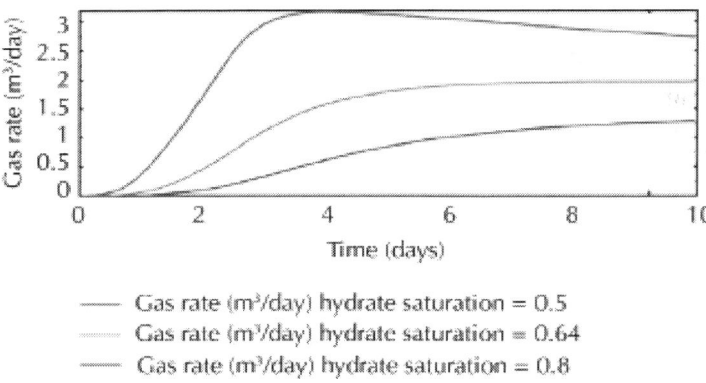

—— Gas rate (m³/day) hydrate saturation = 0.5
······ Gas rate (m³/day) hydrate saturation = 0.64
- - - Gas rate (m³/day) hydrate saturation = 0.8

Figure 9: Effects of initial hydrate saturation on gas production rate.

For a constant temperature of the heat source, a higher saturation of hydrate is expected to correspond to larger volumes of produced CH_4 because of larger hydrate abundance. The substantial increase in the volume of the released gas when hydrate saturation increases from 0.5 to 0.8 as shown in Figure 9 confirms this expectation. Higher saturations of hydrate mean richer sediment and higher production rates. The effect of gas production rate on the heat source temperature is shown below in Figure 10.

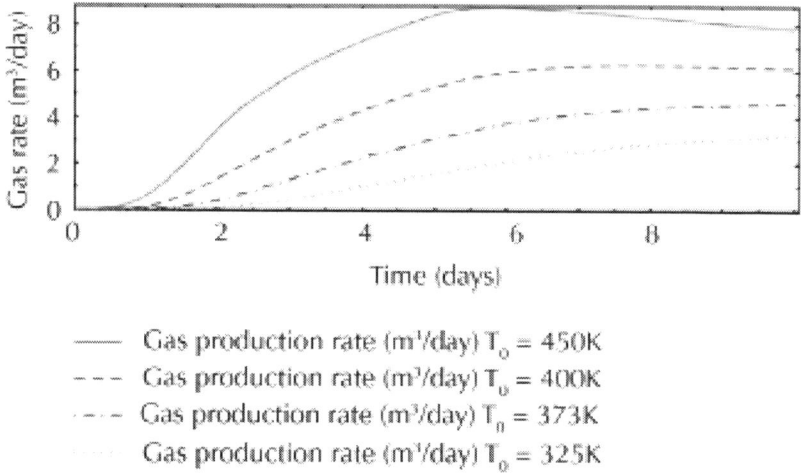

— Gas production rate (m³/day) T_o = 450K
--- Gas production rate (m³/day) T_o = 400K
-·- Gas production rate (m³/day) T_o = 373K
— Gas production rate (m³/day) T_o = 325K

Figure 10: Effect of heat source temperature upon the gas production rate.

The heat source temperature has a marked influence upon the rate of production of gas, with a factor of ~50% increase in production with 12.5% increase in temperature. But maintaining the heat source temperature at say 450 K is a costly proposition. A more economically sustainable temperature would be around 350–360 K.

Validation of Model

The three soil models were validated with the well data from JAPEX/JNOC/GSC and others, Mallik 5L-38, gas hydrate production research well [18] as shown in Figures 11 and 12.

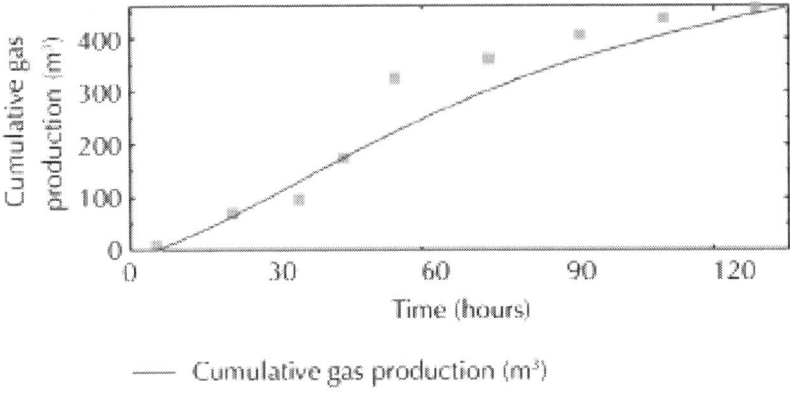

Figure 11: Poroelastic model of sediment validated with Mallik 5L-38 well data.

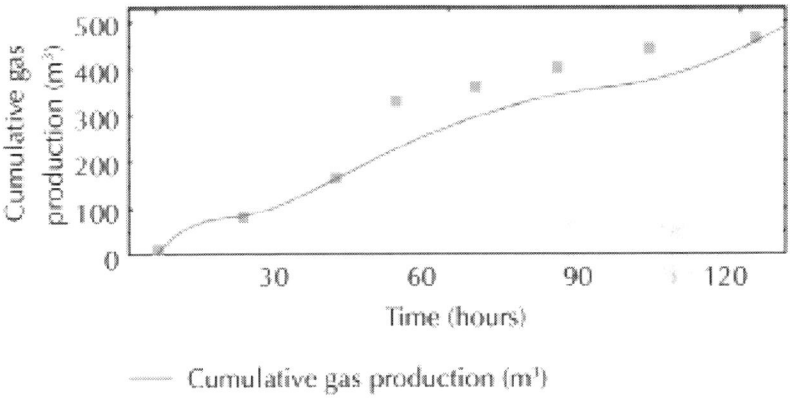

Figure 12: Viscoplastic model of sediment validated with Mallik 5L-38 well data.

FEM Profiles

Figures 13, 14, 15, 16, and 17 show the raw Abaqus screen output for parameters such as nodal temperature, temperature, and porosity across the width and depth of sediment.

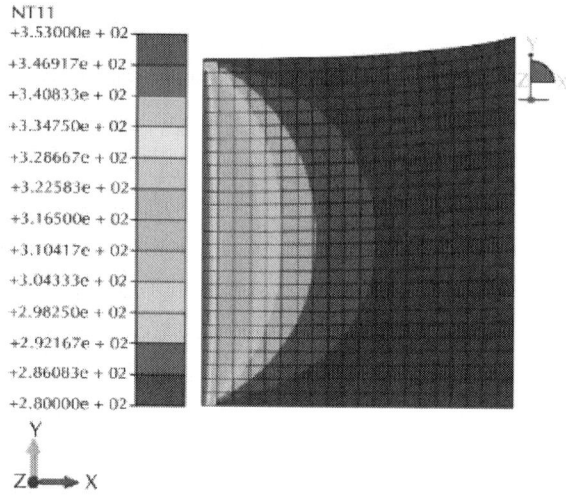

Figure 13: Nodal temperature profile of the sediment bed (with deformation). Radius of bed is 10 m, and depth is 13 m. Initial temperature was 280 K.

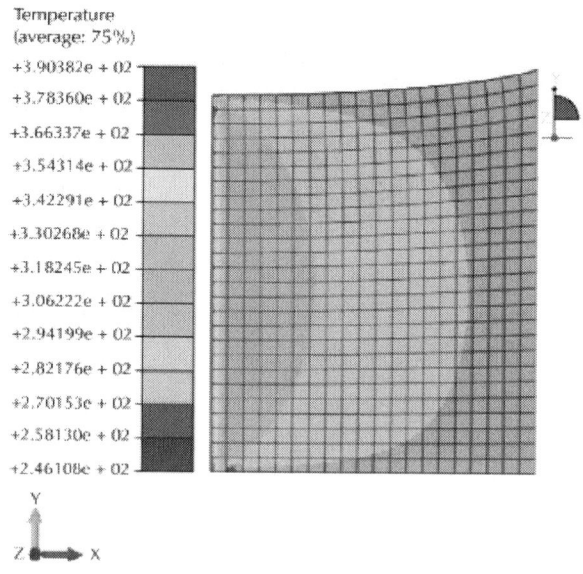

Figure 14: Temperature distribution after 50 days (with deformation).

Effect of Bed Deformation on Natural Gas Production from Hydrates

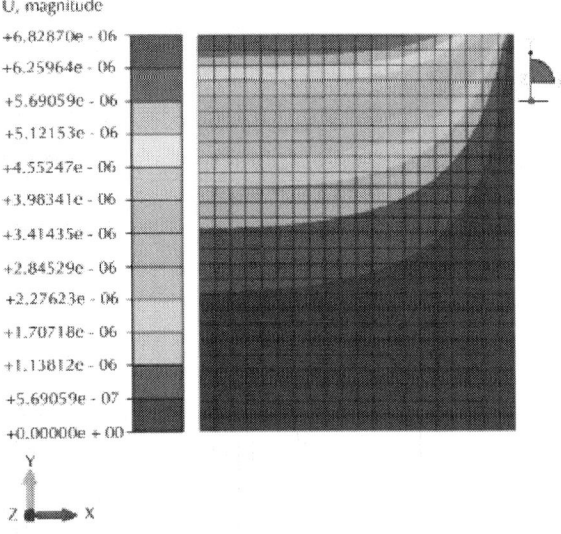

Figure 15: Magnitude of deformation (in m) after 6 hours.

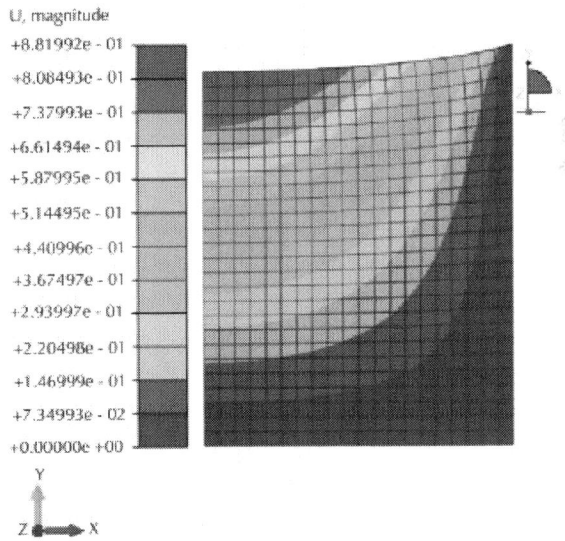

Figure 16: Magnitude of deformation after 50 days (m).

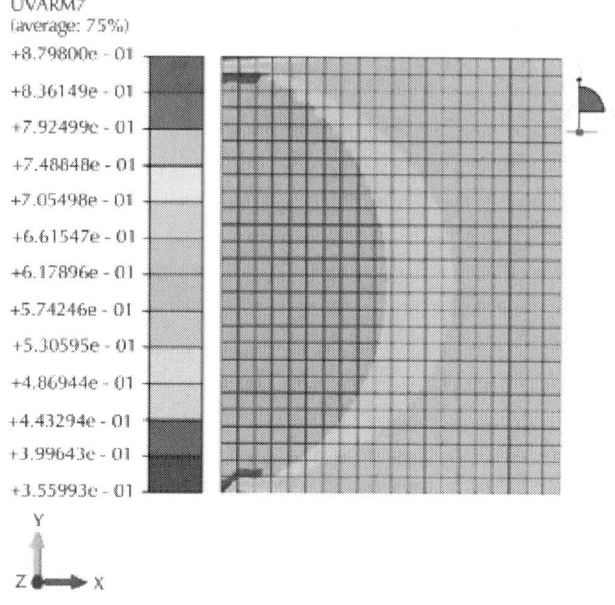

Figure 17: Porosity of sediment after 50 days.

CONCLUSIONS

A commercial package Abaqus was customized to simulate the gas production from natural gas hydrate by considering the deformation of submarine bed. The effect of sediment deformation on gas production by thermal stimulation is studied. The effect of three soil dynamic models, which are generally used for submarine sediments, on the prediction of the gas production has been studied. Gas production rate is found to increase with an increase in the source temperature. Porosity of the sediment and saturation of the hydrate both have been found to significantly influence the rate of gas production.

The economics of the method of gas production from gas hydrates using thermal stimulation has been evaluated in the literature prior to this work. But such evaluations (Table 1) have always assumed that the gas production rate cannot be increased significantly by raising the temperature of the heating fluid. This study has found evidence to the contrary (Figure 10) which, though not conclusive, opens up new possibilities in the economic extraction of natural gas from ocean gas hydrate deposits worldwide in general and the KG and KK basins in India in particular.

Table 1: Economics of gas production from hydrates (modified from [19])

	Gas extraction methods		
	Thermal stimulation	Depressurization	Conventional gas production
Investment in millions of ₹*	254200	166000	157500
Annual cost in millions of ₹	160000	125500	100000
Total production (million m3/year)	1274.26	1557.43	1557.43
Production cost (₹/million m3)	6356.6	4025.9	3213.6
Break-even wellhead price (₹/million m3)	7945.8	5032.3	3972.9

*Calculated for currency rates at 2004 levels.

The higher overall annual average temperature of the regions where these basins are located do mitigate to some extent the energy overhead required to maintain a high temperature flow to the sediment bed.

The earlier techniques of hot fluid injection have many limitations as compared to the method studied in this work. In the earlier method, the heat losses to adjacent sediments are too big to permit economic extraction of gas from hydrate reservoirs by steam injection, even if the steam can be introduced at high rates (15 MW) into impenetrable hydrate reservoirs.

So also, low injection temperatures involve very big volumetric flow rates to carry useful amounts of heat into the reservoir. Injection of approximately 3600 m^3 per day of 66°C water is required if a heat flux of 15 MW is to be maintained. The limitations of disproportionate heat losses on the one hand and unrealistically high injection flow rates on the other hand will probably limit injection temperatures to between 66 and 120°C.

Similarly, unless the porosity is at least 0.15, the heat lost in raising the soil matrix temperature will render thermal stimulation (by injection) ineffective in producing useful quantities of gas. The relative importance of porosity in determining gas production is illustrated in Figure 8.

Scope for future work could be as follows.
- Consideration of inhomogeneity and anisotropy of the hydrate bed.
- Study of the interaction between the multiphase flow and the porous bed.
- Study of hybrid and alternate techniques for gas production.
- Further experimental validation of the above simulated results

REFERENCES

1. "DGH Annual Activity Report 2010-2011," Hydrocarbon exploration and production activities.
2. Y. F. Makogon, "Natural gas hydrates: a promising source of energy," Journal of Natural Gas Science and Engineering, vol. 2, no. 1, pp. 49–59, 2010.
3. M. H. Yousif, H. H. Abass, M. S. Selim, and E. D. Sloan, "Experimental and theoretical investigation of methane gas hydrate dissociation in porous media," in Proceedings of the SPE Annual Technical Conference & Exhibition, pp. 571–18320, October 1988.
4. M. S. Selim and E. D. Sloan, "Heat and mass transfer during the dissociation of hydrate in porous media,"AIChE Journal, vol. 35, pp. 1049–1052, 1989.
5. G. G. Tsypkin, "Mathematical model for dissociation of gas hydrates coexisting with gas in strata,"Doklady Physics, vol. 46, no. 11, pp. 806–809, 2001.
6. C. Ji, G. Ahmadi, and D. H. Smith, "Natural gas production from hydrate decomposition by depressurization," Chemical Engineering Science, vol. 56, no. 20, pp. 5801–5814, 2001.

7. Y. Bai and Q. Li, "Simulation of gas production from hydrate reservoir by the combination of warm water flooding and depressurization," Science China Technological Sciences, vol. 53, no. 9, pp. 2469–2476, 2010.
8. X. Sun, N. Nanchary, and K. K. Mohanty, "1-D modeling of hydrate depressurization in porous media,"Transport in Porous Media, vol. 58, no. 3, pp. 315–338, 2005.
9. S. Kimoto, F. Oka, T. Fushita, and M. Fujiwaki, "A chemo-thermo-mechanically coupled numerical simulation of the subsurface ground deformations due to methane hydrate dissociation," Computers and Geotechnics, vol. 34, no. 4, pp. 216–228, 2007.
10. S. Kimoto, F. Oka, and T. Fushita, "A chemo-thermo-mechanically coupled analysis of ground deformation induced by gas hydrate dissociation," International Journal of Mechanical Sciences, vol. 52, no. 2, pp. 365–376, 2010.
11. A. Vysniauskas and P. R. Bishnoi, "A kinetic study of methane hydrate formation," Chemical Engineering Science, vol. 38, no. 7, pp. 1061–1072, 1983.
12. P. Englezos, N. Kalogerakis, P. D. Dholabhai, and P. R. Bishnoi, "Kinetics of formation of methane and ethane gas hydrates," Chemical Engineering Science, vol. 42, no. 11, pp. 2647–2658, 1987.
13. H. C. Kim, P. R. Bishnoi, R. A. Heidemann, and S. S. H. Rizvi, "Kinetics of methane hydrate decomposition," Chemical Engineering Science, vol. 42, no. 7, pp. 1645–1653, 1987.
14. J. S. Pic, J. M. Herri, and M. Cournil, "Experimental influence of kinetic inhibitors on methane hydrate particle size distribution during batch crystallization in water," Canadian Journal of Chemical Engineering, vol. 79, no. 3, pp. 374–383, 2001.
15. P. D. Dholabhai, N. Kalogerakis, and P. R. Bishnoi, "Kinetics of methane hydrate formation in aqueous electrolyte solutions," Canadian Journal of Chemical Engineering, vol. 71, no. 1, pp. 68–74, 1993.
16. N. Gnanendran and R. Amin, "Modelling hydrate formation kinetics of a hydrate promoter-water-natural gas system in a semi-batch spray reactor," Chemical Engineering Science, vol. 59, no. 18, pp. 3849–3863, 2004.

17. D. Kashchiev and A. Firoozabadi, "Driving force for crystallization of gas hydrates," Journal of Crystal Growth, vol. 241, no. 1-2, pp. 220–230, 2002.
18. S. H. Hancock, T. S. Collett, and S. R. Dallimore, "Overview of thermal-stimulation production test results for the JAPEX/JNOC/GSC et al. Mallik 5L-38 gas hydrate production research well," 2005.
19. B. S. Pierce and T. S. Collett, "Energy resource potential of natural gas hydrates," in Proceedings of the 5th Conference Exposition on Petroleum Geophysics, pp. 899–903, Hyderabad, India, 2004.

Chapter 3

Thermochemical Equilibrium Model of Synthetic Natural Gas Production from Coal Gasification Using Aspen Plus

Rolando Barrera[1], Carlos Salazar[2], and Juan F. Pérez[3]

[1]Grupo CERES, Departamento de Ingeniería Química, Facultad de Ingeniería, Universidad de Antioquia (UdeA), Calle 70 No. 52-21, Medellín, Colombia

[2]Celsia S.A. ESP, Sede zona Franca Celsia, vía 40 No. 85-555, Barranquilla, Colombia

[3]Grupo de Manejo Eficiente de la Energía (Gimel), Departamento de Ingeniería Mecánica, Facultad de Ingeniería, Universidad de Antioquia (UdeA), Calle 70 No. 52-21, Medellín, Colombia

ABSTRACT

The production of synthetic or substitute natural gas (SNG) from coal is a process of interest in Colombia where the reserves-to-production ratio (R/P) for natural gas is expected to be between 7 and 10 years, while the R/P for coal is forecasted to be around 90 years. In this work, the process to produce SNG by means of coal-entrained flow gasifiers is modeled under thermochemical equilibrium with the Gibbs free energy approach. The model was developed using a complete and comprehensive Aspen Plus model. Two typical technologies used in entrained flow gasifiers such as coal dry and coal slurry are modeled and simulated. Emphasis is put on interactions between the fuel feeding technology and selected energy output parameters of coal-SNG process, that is, energy efficiencies, power, and SNG quality. It was found that coal rank does not significantly affect energy indicators such as cold gas, process, and global efficiencies. However, feeding technology clearly has an effect on the process due to the gasifying agent. Simulations results are compared against available technical data with good accuracy. Thus, the proposed model is considered as a versatile and useful computational tool to study and optimize the coal to SNG process.

INTRODUCTION

Coal is a major source of energy, accounting for ~25% of the world energy supplies and ~40% of the world electricity generation. It is predicted that coal will continue to play an important role in meeting the world's increasing energy demands in the foreseeable future [1]. However, the use of coal faces several challenges such as clean and efficient energy systems, the challenge of carbon storage and sequestration, and the environmental impacts due to the mining [1, 2]. The coal gasification in entrained flow reactors with steam and/or oxygen produces synthesis gas (syngas), which is a mixture of mostly hydrogen and carbon monoxide [3]. Therefore, the urgent needs to produce fuels and chemical products from syngas prompt the study of this thermochemical process [3, 4]. Gasification units in electric power plants produce a fuel gas to drive gas turbines. And gasification in chemical plants yields syngas that can be used to produce a wide spectrum of chemical products, such as ammonia, methanol, methane, and liquid fuels [5]. Future plants will be hybrid power/chemical plants with gasification as the key unit operation; as a consequence, the thermochemical process has been emerging as the premier unit operation in the energy and chemical industries. Therefore, the gasification continues to be an important topic to research [3, 6, 7].

The production of SNG from coal is a process of particular interest to Colombia, where depletion of natural gas is being foreseen in the coming years due to the increasingly high level of demand of the last several years [8]; carbon reserves are still foreseen to last for around 90 years [9, 10]. These reserves are the highest coal reserves in South America. The coal production in Colombia is distributed by regions as follows: 50.92% Cesar, the main coal mines are El Descanso, Calenturitas, La Loma, and Jagua; 38.87% La Guajira (Cerrejón coal mine); 3.57% Cundinamarca (Guaduas coal mine); 3.21% Boyacá (Tunja-Duitama, Sogamoso-Jericó, and Chinavita-Umbita-Tinabá coal mines); 2.22% Norte de Santander (Zulia, Cúcuta, Tasajero, and Toledo coal mines); and 1.21% others. Therefore, one of the main challenges facing the country is finding how to add value to mineral resources under environmental, efficiency, and sustainability criteria [1]. Therefore, this study is focused on modeling the SNG production process by means of coal gasification in entrained flow as a promising and clean alternative way to produce energy and fuels.

Several studies of modeling coal gasification have been published in the specialized literature. Gräbner and Meyer [11] analyzed the gasification process under the first and second thermodynamic laws. The aim has been to study the coal rank effect (standard and high ash content coals) on the gasifier technology (Shell, Siemens, Texaco, ConocoPhillips, and High Temperature Winkler-HTW). The higher exergy efficiencies were reached by slurry feeding and dry feeding technologies with the standard coal.

Kunze and Spliethoff [12] developed a model with Aspen Plus to simulate a generic entrained flow gasifier. The objective was to analyze the effect of the fuel feed system, that is, dry feed and slurry feed on the gasification process at 30 bar. They found higher energy efficiency for dry feed gasification (83%) when compared with slurry feed (72%). Seifkar et al. [13] studied the effect of coal supply and reactor cooling system on the entrained flow gasifier process. They analyzed three systems: the first one included dry coal feed and reactor cooling with water, the second consisted of one dry coal feed with partial cooling with water, and the last one was a coal slurry feed supply system without refrigeration. The authors discussed advantages and disadvantages associated with the studied systems.

Maurstad et al. [14] presented a model with Aspen Plus to characterize an integrated gasification combined cycle plant (IGCC) with and without CO_2 capture. They carried out a comparison between two technologies (dry feed and slurry feed) with five types of coal in the IGCC plant energy behavior. They found that the thermal efficiency and power of the IGCC plant diminished with the coal rank for the slurry feed technology, while the dry feed technology was not affected. Yu et al. [15] studied the effect of the gasification technology on the water gas shift reaction unit used for Fischer-Tropsch processes. The study was conducted by means of an Aspen Plus model. They showed that the dry feed technology presented higher energy efficiency and lesser H_2/CO ratio with regards to the slurry feed technology. Prins et al. [16] studied different ways of carbon capture and sequestration on an IGCC plant by process modeling. The dry feed technology was analyzed because of its higher efficiency in the gasification process (82.3%) against the gasifiers with slurry feeding systems (about 75%).

Bockelie et al. [17] studied the syngas composition and the cold gas efficiency of two commercial gasifiers using a computational fluids dynamic model (CFD). The commercial gasifiers considered were Shell (dry feed technology with one gasification stage) and ConocoPhillips (slurry feed technology with two gasification stages). The authors highlight the agreement between the simulation results with the proposed model and the data reported in the literature. Armin [18] and Silaen and Wang [19] conducted numerical simulations of the gasification process in entrained flow reactors. They studied the effect of the operating parameters such as coal dry feeding and slurry feeding systems, type of gasifying agent, and coal rank in the global gasifier performance. They highlighted that syngas heating value (HHV_{syngas}) is higher with the dry system technology. Moreover, the carbon conversion efficiency and HHV_{syngas} increase when oxygen is used as gasifying agent because the nitrogen inert effects of the air are avoided.

Using biomass as a feedstock to produce SNG by means of gasification, Vitasari et al. [20] developed an Aspen Plus model to conduct an exergy analysis of different biomass types (wood, urban solid wastes (USW), and sewage sludge). They simulated the process under different operating conditions, including variations in the reactor pressure as well as temperature and pressure in the methanation reactor. The global process exergy efficiency was reached with wood being between 53 and 58%; for USW was between 42 and 46% and for the sewage sludge being from 47 to 57%. Tremel et al. [21] simulated a small scale plant to produce SNG via indirect steam biomass gasification. The thermal biomass input to the gasifier was 500 kW. The model was developed under equilibrium approach using Aspen Plus and considering the minimization of the Gibbs free energy. They studied the effect of four operation conditions on the energy efficiency of the process. The ratio between the energies of SNG and biomass reached values around 66–75%. This indicator increases when the pyrolyzed char is used to feed the steam boiler.

Heyne et al. [22] studied the thermal integration between an existing biomass steam and power cycle (CHP) and the production process of SNG via indirect gasification. The thermal plant was modeled and simulated by means of Aspen Plus, and the analysis was conducted using pinch technology. The process global efficiency reached was about 90%, and it indicates that the SNG production is feasible via indirect gasification coupled to a CHP plant.

On the other hand, the Department of Energy from the United States through the National Energy Technology Laboratory conducted a diagnosis of commercial entrained flow gasifiers. The analysis was carried out by means of Aspen Plus models and validated with data from 14 IGCC plants. The IGCC plants efficiencies without CO_2 capture were 42.1%, 39.7%, and 39% for Shell, ConocoPhillips, and Texaco, respectively. Regarding the costs, the specific costs of the IGCC plants without CO_2 capture were about 2350 \$USD/kW for ConocoPhillips (slurry feed technology) and 2710 \$USD/kW for Shell (dry feed technology) [23].

The coal gasification process has been widely studied under different operational conditions. This has been done by means of modeling and simulation strategies with software tools such as Aspen Plus to analyze the thermochemical process or to study how the input parameters affect the IGCCs process. According to the literature cited, the effect of coal gasification technology on the SNG production using

Aspen Plus model has not been presented. Therefore, in the current study, two typical technologies used in entrained flow gasifiers with dry coal and slurry coal feed systems are modeled and simulated. Emphasis is put on interactions between the fuel supply technologies and energy output parameters of coal-SNG process, including carbon conversion efficiency, cold gas efficiency, process and global energy efficiencies, SNG heating value, Wobbe Index, and power. Since most of the revised literature did not go deep into model details, this study presents as an additional contribution, a comprehensive and complete model of coal to SNG process under thermochemical equilibrium developed in Aspen Plus.

METHODOLOGY

Coal-SNG Process Description: coal Slurry Feed Mode

The complete process to produce SNG via coal gasification in entrained flow reactors with slurry feeding system includes several stages that are shown in Figure 1. Stage 1 includes the solid fuel inlet into the system as a slurry after mixing it with water. The typical concentration of water in the slurry is around 30–40% wt., which means that a significant portion of heat generated by coal combustion will be used to vaporize water in the gasifier at higher pressure. Therefore, the cold gas efficiency (CGE) of wet feed gasifiers is expected to be lower (between 8 and 10%) than the dry feed gasifiers [12, 16].

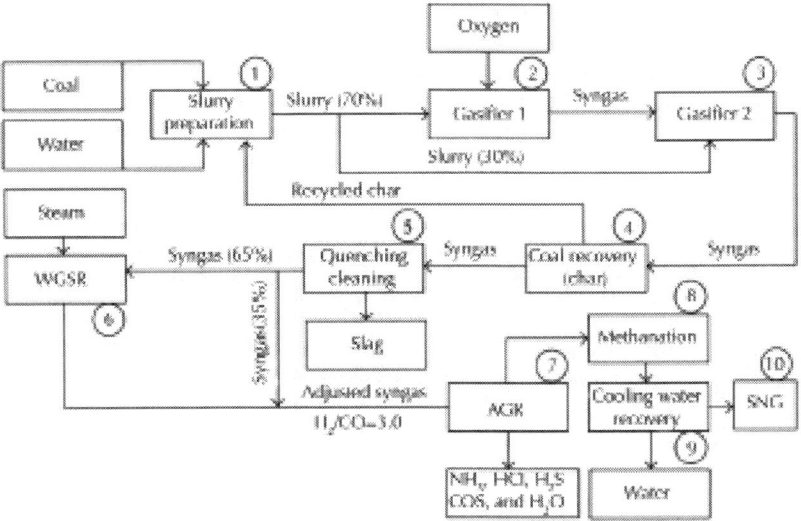

Figure 1: Process stages of SNG production via coal gasification using slurry feeding technology.

The entrained flow gasifier is modeled with two gasification stages, 70% of slurry mass flow goes into a pressurized reactor (Stage 2, Figure 1) where the slurry reacts with oxygen (98% purity). The use of air as an oxidant is avoided due to high flow rates that produce stack gas by the higher amount of nitrogen. The syngas formed in this reactor flows to a second pressurized reactor (Stage 3, Figure 1) where reaction takes place with the remaining 30% of the initial slurry mass flow. In the second reactor, there is not an addition of oxygen. In this way, the coal devolatilization is promoted and the reduction reactions for CO_2, H_2, and H_2O with char are driving to formation of H_2, CH_4, and CO as majority compounds, (1)–(3). Consider

$$C + CO_2 \Longleftrightarrow 2CO \tag{1}$$

$$C + 2H_2 \Longleftrightarrow CH_4 \tag{2}$$

$$C + H_2O \Longleftrightarrow CO + H_2 \tag{3}$$

The units downstream from the gasifier are mostly standard gas-phase processes [3]. Syngas produced in the gasification process (Stage 23, Figure 1) goes through a separator that removes unreacted coal (Stage 4, Figure 1). Unreacted coal is reentered to the process as recycled char in Stage 1. Further, in Stage 5 (Figure 1) the syngas is quenched and cooled with water with the corresponding ash solidification (slag formation). The slag is removed from the syngas in a particulate removal system. According to the particle size distribution, the particulate matter clean-up process can include cyclones, electrostatic precipitators, and bag filters. After cooling and particulate removal, the syngas is fed into a water gas shift reactor (WGSR), where hydrogen formation is promoted (Stage 6, Figure 1). By adding steam to the WGSR, carbon monoxide and water react to carbon dioxide and hydrogen (see (4)). In this way, the WGSR can produce a modified syngas with specific molar ratio H_2/CO by adjusting any of the reactor inlet flows (steam and/or syngas). Consider

$$CO + H_2O \Longleftrightarrow CO_2 + H_2 \tag{4}$$

The molar ratio H_2/CO at the exit of the WGSR will depend on aims of the entire plant process; that is, for dimethyl synthesis, a molar ratio $H_2/CO=1$ is desired, while Fischer-Tropsch process requires a ratio $H_2/CO=2$ For SNG production, a molar relation $H_2/CO=3$ is preferred [21, 24, 25]. Thereby, a portion of the syngas from Stage 5 (Figure 1) is bypassed and only between 60% and 70% of the syngas mass flow goes through the WGSR. In chemical applications, the synthesis gas and/or hydrogen is fed to downstream chemical plants. The carbon dioxide is suitable for sequestration [3], or it can be also used as a carrier gas in entrained flow gasifiers with dry feeding technologies [12]. In the SNG production scheme (Figure 1), the syngas with adjusted-molar ratio H_2/CO is then passed through an acid gases recovery unit (AGR), which consists of a separation unit that removes acid gases (Stage 7, Figure 1). In the AGR unit the syngas is cleaned up from acid gases and other impurities that are formed upstream in the process, such as NH_3, HCl, H_2S, COS, H_2O, and CO_2, among others. This cleaning process is necessary because the next process stage (Stage 8, Figure 1) consists of a catalytic reactor whose performance highly depends on the quality of the reactor inlet gases [24]. Furthermore, in the reducing environment present in gasifiers, the sulfur and nitrogen impurities appear as hydrogen sulfide and ammonia, respectively. Both of these chemicals can be easily removed using pollutant removal systems (i.e., sulfur dioxide absorbers and NOx reactors) and are potentially valuable by-products to be used as well or as raw material for different chemical process, that is, a sulfuric acid plant production. In the current process, the outcoming gas from Stage 7 (Figure 1) is constituted mainly by H_2 and CO. This syngas with adjusted-molar ratio $H_2/CO=3$ flows to the methanation stage (Stage 8, Figure1) where methane formation is promoted (see (5)). Even when additional reactions could occur in the methanation reactor, that is, (6), the methane formation by (5) is thermodynamically favored due the operating conditions in the reactor (temperature and pressure) as well as the composition of the reactor inlet gases and the adapted catalytic technology. Simulations in this study revealed that, under the conditions given, the equilibrium constant of the reaction presented in (6) is 22 times lower than equilibrium constant of the reaction presented in (5). Thus, there are no significant errors in assuming (5) as the only reaction taking place on the methanation reactor.

Stage 9 (Figure 1) corresponds to the SNG conditioning stage. The SNG is cooled; therefore, steam formed in Stage 8 (see (5)) is condensed and separated from the SNG. The outlet of this block (Stage 10, Figure 1) corresponds to the desired natural gas at temperatures around 50–60°C with high level concentrations of methane (SNG). Consider

$$CO + 3H_2 \Longleftrightarrow CH_4 + H_2O \quad (5)$$

$$CO_2 + 4H_2 \Longleftrightarrow CH_4 + 2H_2O \quad (6)$$

Coal-SNG Process Description: dry Coal Feed Mode

The process diagram to produce SNG via coal gasification using an entrained flow reactor with dry feeding technology is presented in Figure 2. Stage 1 represents the fuel inlet to the system mixed with CO_2 in an entrained flow gasifier. It should be noted that compared with the slurry feeding technology, this process require that the coal goes to the gasifier with very low moisture content, between 1% and 2%. In the dry feeding technology, the gasifier includes just one reactor [26] (Stage 2, Figure 2). All remaining stages in the dry feeding technology correspond to the analogous stages that were described in the slurry feeding technology (Section 2.1). However, the amount of syngas that passes through the WGSR is different for both technologies. In the dry feeding technology, around 85%–95% of the syngas mass flow goes through the WGRS and the remainder of the syngas is bypassed to acid gases recovery unit (AGR). This is because the syngas molar ratio CO/H_2 leaving the gasifier in slurry technology is around 1:1, while in dry technology the syngas CO/H_2 molar ratio is around 2:1. The higher CO concentration in the syngas leaving from the gasifier with dry feeding technology is attributed to the lower amount of water in the gasification process, which is directly related with the H_2 concentration.

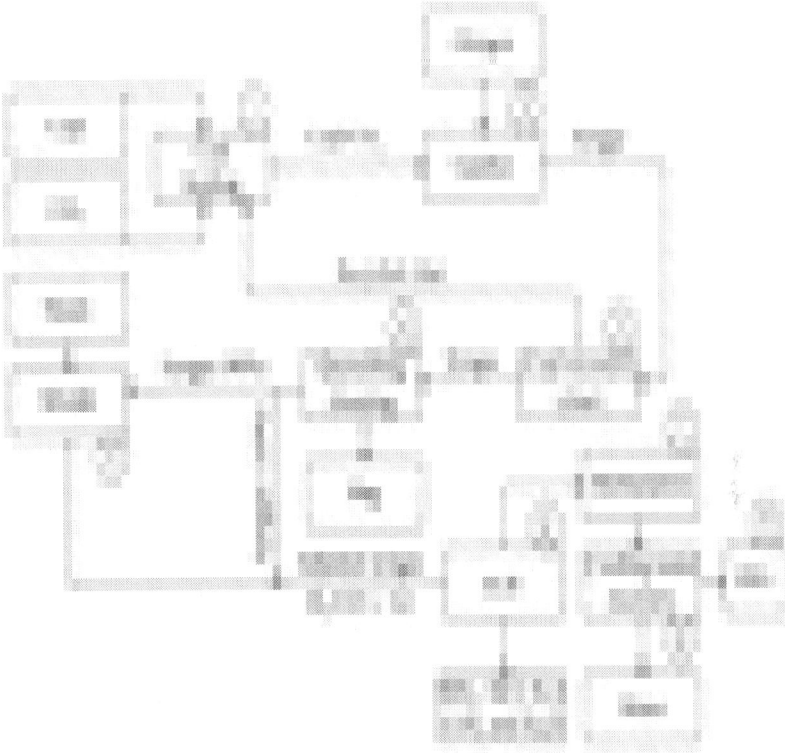

Figure 2: Process stages of SNG production via coal gasification using dry feeding technology.

Test Fuels

The chemical characterization of the Colombian coals used in this study is presented in Table1. According to the ASTMD88-12 standard, the Sanoha coal is classified as bituminous coal and Bijao coal is a subbituminous type B coal. These fuels are used to validate the model with technical data.

Table 1: Chemical characterization of Colombian coals (Bijao coal and Sanoha coal) used in simulations

Characterization	Bijao	Sanoha
Proximate analysis (wt.% wet)		
Ash (ASTM D 73174)	6.5	14.8
Volatile matter (ASTM D 3175)	45.89	29.85
Fixed carbon (ASTM D 3172)	47.61	55.35
Moisture content (ASTM D 3302)	19.05	4.93
Ultimate analysis (wt. % dry)		
C	68.24	72.11
H	4.9	4.78
N	1.59	1.62
S (ASTM D 5865)	1.36	1.44
O	17.38	5.52
Cl	0.03	0
Ash (ASTM D 73174)	6.5	14.8
HHVd.b (kcal/kg) (ASTM D 5865)	5407	7538

Software Selection

Simulations were carried out using Advanced System for Process Engineering simulation software (Aspen Plus v7.3) [27]. Aspen Plus is a process modeling software suitable for a variety of steady state modeling applications. Currently, this software was widely applied in simulating gasification processes, cogeneration plants, and polygeneration systems, all of them with different technologies and fuels [12, 28–34], and good agreement between the industrial data and those determined using the Aspen models was obtained [35, 36]. Aspen Plus software provides a flexible input language for describing the SNG production process, including its components, connectivity, and computational sequences. Use of Aspen Plus leads to an easier way of model development, maintenance, and updating since small sections of complex and integrated systems can be created and tested as separate modules before they are integrated. It has an extensive physical properties database to model the material streams in SNG production process [6, 30, 33, 37, 38]. Additionally, Aspen Plus has many built-in model blocks (such as heaters, pumps, stream mixers, and stream splitters), some of which can be directly used in this work.

ASPEN PLUS MODEL DESCRIPTION

Hypothesis

The relatively high temperatures used in the gasification process allow the consideration that the kinetic barriers are minimized and it was found that the gaseous mixtures leave from the gasifiers approach to the equilibrium [4, 39]. Therefore, the gasification process can be successfully described by means of a thermodynamic model [27, 40]. In this work, an overall equilibrium approach is employed while neglecting the hydrodynamic complexity of the gasifier. Other assumptions in the model are: (1) stationary state, (2) WGSR and methanation reactor which are equilibrium-isothermal reactors, (3) char which is supposed to be 100% graphite (conventional substance available in Aspen Plus database), (4) the ash content which is turned into slag, and (5) the gasifier reactors which are simulated as RGibbs adiabatic reactors. In addition, the conditioning coal (drying) and air separation unit to get oxygen as a gasifying agent are not considered in the model.

The RGibbs reactor was chosen because this kind of reactor can handle three phases under chemical equilibrium and allows predicting the equilibrium composition of the produced syngas by minimizing the Gibbs free energy [21]. Indeed, the RGibbs reactor of Aspen that works under the Gibbs free energy minimization principle has been widely adopted to represent gasification reactions [6, 41–43]. The equilibrium products potentially formed in the RGibbs reactors are H_2O, N_2, O_2, H_2, C, CO, CO_2, CH_4, H_2S, NH_3, COS, HCl, and Cl_2 [24,28]. These species allow the versatile simulation of different kind of syngas, depending on the input parameters. Definition of species potentially formed in the gasifier was based on the following considerations. (i) CH_4 is the only hydrocarbon taken into consideration in this work due to higher pressure and temperature [4]. (ii) The sulfur contained in the coal is assumed to be converted mainly into H_2S and COS. The low amounts of chlorine suggest that the chlorinated species formed are only traces of HCl and Cl_2; and (iii) the assumption that only NH_3 forms and not oxides of nitrogen are produced has already been made by other researchers [44].

Coal to SNG Model

This section discusses how Aspen Plus is used to simulate the SNG production via coal gasification using slurry feeding technology. Also, the main differences are described in the modeling approach for dry feeding technology. The Aspen Plus process flow sheet is divided into five hierarchies (Figure 3): slurry preparation (FEEDING), gasification (GASIFIC), water gas shift reactor (SHIFTING), acid gas recovery unit (AGR), and methanation (METHANAT). Each one of these hierarchies includes at least one Aspen Plus built-in block, and in some cases, that is, in the FEEDING hierarchy, there are also included additional Fortran subroutines.

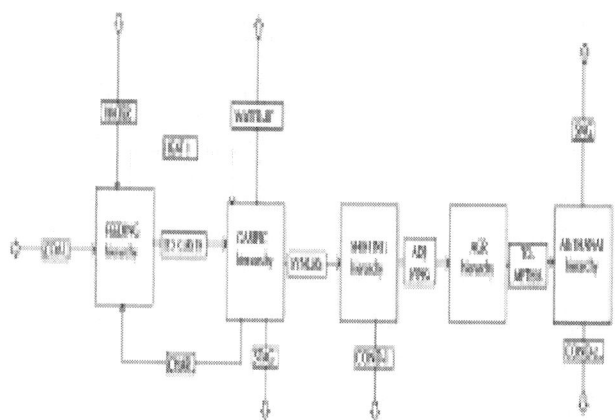

Figure 3: Aspen Plus process flow sheet for SNG production via coal gasification using slurry feeding technology (ConocoPhillips 66).

In the case of dry feeding technology, the main process diagram is quite similar to the one shown in Figure 3. In the flow sheet of the dry feeding technology, the WATER inlet stream is replaced by a CO_2 inlet stream. Differences between the models of both feeding modes are appreciated inside some specific hierarchies that are described below.

FEEDING Hierarchy

The Aspen Plus model used for simulating the slurry preparation (FEEDING hierarchy) is showed in Figure 4.

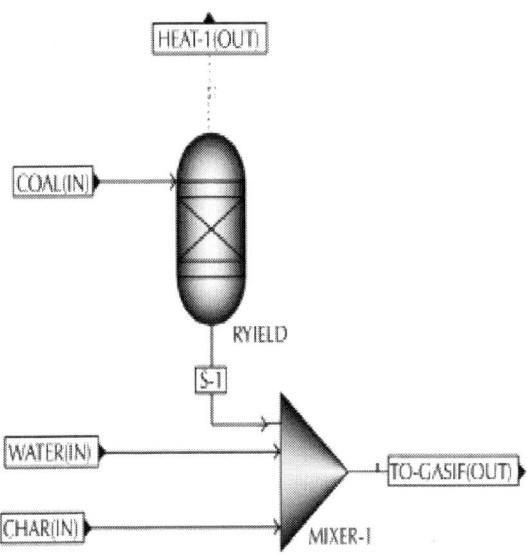

Figure 4: Aspen Plus feeding hierarchy for the simulation of the slurry preparation in the SNG production via coal gasification.

FEEDING hierarchy (Figure 4) is used to simulate the raw material inlet to the process (Stream COAL) which is composed only by coal as received in the plant; that is, neither crushing processes nor drying processes are considered. The feedstock is Colombian coal with the proximate analysis and ultimate analysis given in Table 1. Aspen Plus cannot handle nonconventional substances, and coal is a nonconventional solid with a complex macromolecular structure [45, 46]. Therefore, the coal stream needs to be hypothetically decomposed in reactive compounds, that is, its corresponding constituents (such as C, H_2, N_2, O_2, S, Cl_2, and H_2O), based on its proximate analysis and ultimate analysis. This is performed in a yield reactor (RYIELD block, Figure 4) [41, 43–46]. The yield distribution for this reactor has been specified by FORTRAN statements in a calculator block [47]. These statements specify the mass flow rates of the components in the stream S-1, Figure 4. The slurry is then formed by mixing the stream S-1 with the inlet stream WATER in the mixer block MIXER-1, and it is driven to the next process stage through the stream TO-GASIF (Figure 4). The stream TO-GASIF leaves from the FEEDING hierarchy and becomes the main inlet to the GASIFIC hierarchy, where the formation, quenching, conditioning, and cleaning of the syngas are simulated. In the built-in mixer block (MIXER-1), the unreacted solid coal (char) from the gasification process stage is reentered to the process. By the other side regarding the energy, the heat of reaction associated with the coal decomposition is carried by the heat stream HEAT-1 (Figure 4) into the next hierarchy (GASIFIC, Figure 3) [25, 30], where gasification reactions have been modeled [47]. Then, further gasification reactions were applied to the available constituents of the coal at the same enthalpy level.

When dry feeding technology is simulated, the Aspen Plus flow sheet for FEEDING hierarchy is quite similar to that described in Figure 4; the only difference is the use of a CO_2 inlet stream instead of the WATER inlet stream, taking into account the lower moisture content of the raw material (1%-2%).

GASIFIC Hierarchy

Figure 5 is a schematic diagram of the Aspen Plus model for the gasifier facility [48, 49]. The gasification facility consists of three sections: a reactor, quenching, and cleaning systems (solids and liquid removing from the syngas). In the modeled slurry technology, the gasification system is comprised of two RGibbs reactors (RGIBB-1 and RGIBB-2, Figure 5). The stream TO-GASIF coming from FEEDING hierarchy is split into two streams by a stream divisor block (SPLIT1). The formed streams S-2 and S-3 (Figure 5) carry on the 70% wt. and 30% wt. (resp.) of the total material coming in the stream TO-GASIF. In the same way, the heat stream HEAT-1 is divided in two heat streams H-2 and H-3 with a stream divisor block (QSPLIT), Figure 5. The stream S-2 reacts with the inlet stream OXYGEN in the RGIBB-1 reactor. The mass flow of the OXYGEN stream can be adjusted in order to modify the equivalence ratio (ER) (see (7)) or the reactor temperature. The stream

S-4 (Figure 5) feeds the second RGibbs reactor RGIBB-2 and reacts with the stream S-3. In this way, the formation of H_2, CH_4 and CO is promoted; see (1)–(3). The stream S-5 (Figure 5) is then quenched and cooled by mixing it with water from the inlet stream WATQFEED in the Aspen Plus built-in blocks MIXER-2 and HEATER-1. The next step is the block SSPLIT, it is used to simulate a stream divisor. SSPLIT is an Aspen Plus built-in block that allows completely separating conventional from no conventional compounds as well as solids from liquids and gases. Thus, the SSPLIT separates conventional solids and unconventional solids and the gaseous phase coming from the reactor [6, 47]. The conventional solids correspond to the unreacted coal or char (stream CHAR), and the unconventional solids are mainly ash (stream SLAG). Stream CHAR is recycled and fed back into the process in the FEEDING hierarchy; in contrast, the stream SLAG is discharged from the process. The remaining wet gaseous mixture (stream S-8, Figure 5) is then driven to the SEP-1 block where water recovery is simulated and the clean-dry syngas is driven to the next SHIFTING hierarchy. SEP units used in this model are Aspen Plus built-in blocks which separate substances from a mixture by means of mass and energy balances criteria without any thermodynamic equilibrium calculation. In several commercial gasifiers, quenching is used for gas cooling and promotion of slag formation. The water in the WATER-RE stream can be conditioned and reused in the WATQFEED or the WATER inlet streams.

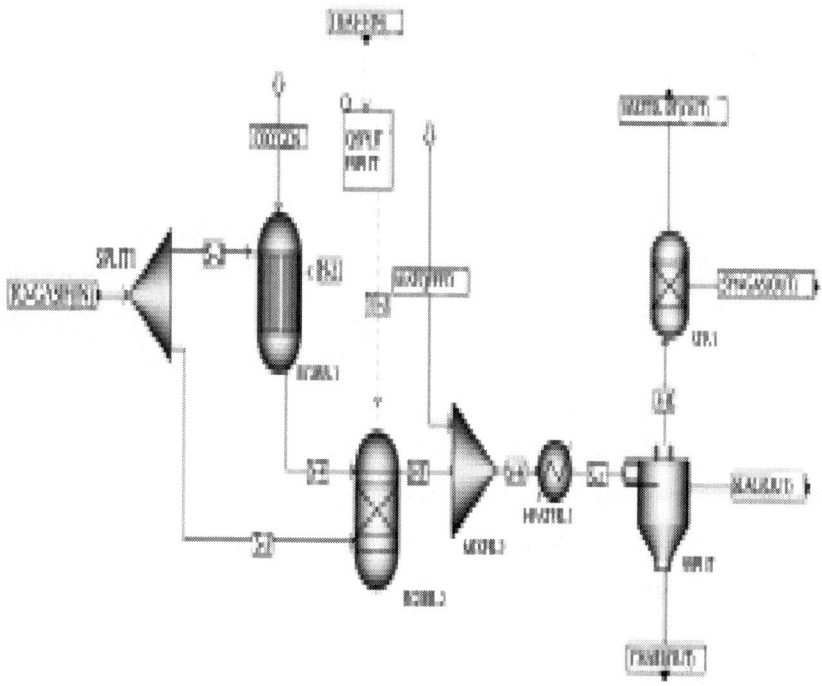

Figure 5: Aspen Plus GASIFIC hierarchy diagram to simulate the gasification stage of coal-SNG process. Flow sheet for the slurry feeding technology.

This developed model is useful to predict the syngas composition and reactor temperature under various operating conditions, including flow rates, composition, and temperature of the feed materials as well as the operating reactor pressure.

To simulate the dry feeding technology, the Aspen Plus flow sheet for GASIFIC hierarchy uses just one RGibbs reactor; hence neither mass splitters nor heat splitters are introduced into the model, and the full TO-GASIF stream as well as the full HEAT-1 stream goes to the RGIBB-1 reactor; see Figure 6. Despite this, all the remaining built-in blocks in the dry feeding technology (Figure 6) correspond to the analogous stages that are described in the slurry feeding technology (Figure 5).

Thermochemical Equilibrium Model of Synthetic Natural Gas ...

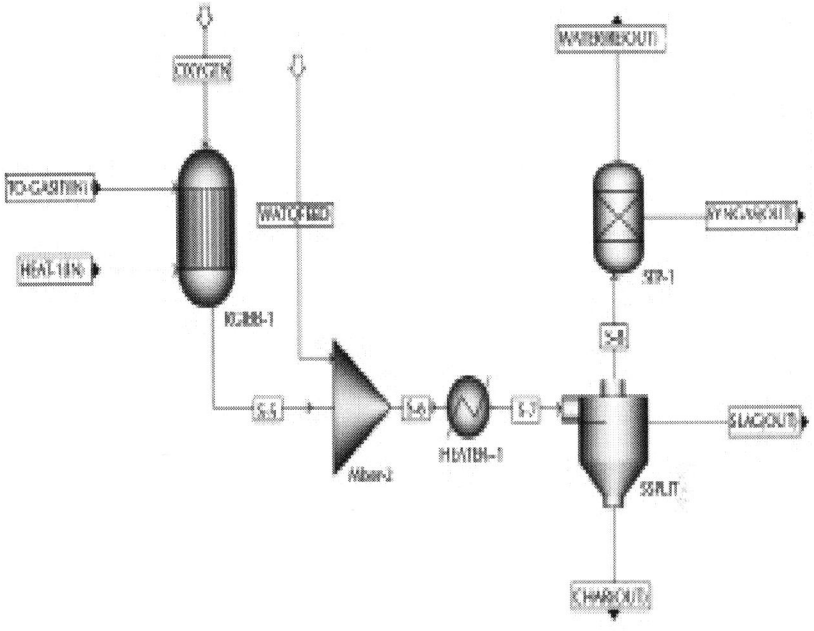

Figure 6: Aspen Plus GASIFIC hierarchy diagram to simulate the gasification stage of coal-GNS process. Flow sheet for dry feeding technology.

SHIFTING Hierarchy

The syngas coming out from GASIFIC hierarchy is sent to the SHIFTING hierarchy, where the WGSR is simulated. As shown in Figure 7, the bypass is simulated by means of a built-in stream splitter block (SPLIT-2). The stream S-9, in the slurry feeding technology, carries between 60% and 70% of SYNGAS stream to the REQUIL reactor. In this reactor, the water gas shift reaction (see (4)) takes place. The stream S-11 leaving the WGSR reactor and stream S-10 (bypass) are mixed and cooled in the HEATER-2 built-in block. The mass flow of the inlet stream STEAM is adjusted until a molar ratio $H_2/CO=3$ in the stream ADJ-SYNG. To adjust this ratio, the sensitivity analysis tool supported by Aspen Plus was used. This tool is in the section model analysis. The stream ADJ-SYNG leaves the SHIFTING hierarchy and feeds the AGR hierarchy, where gas cleaning process is simulated.

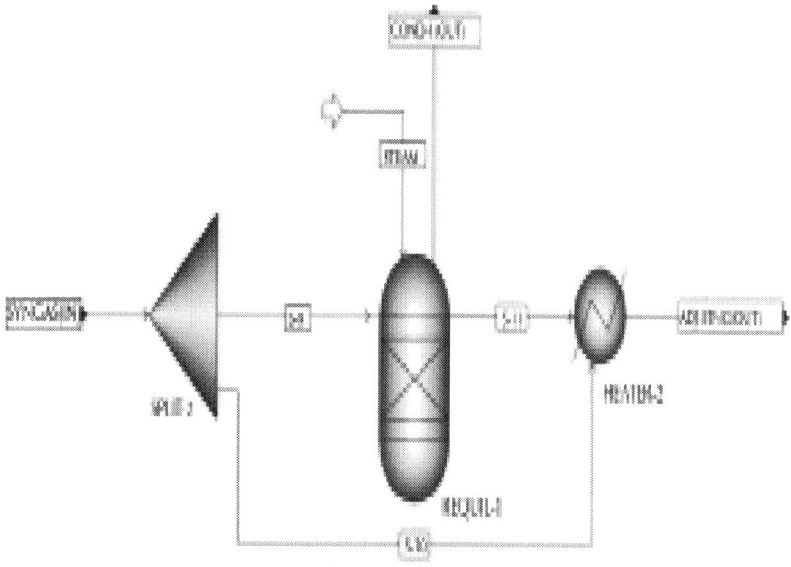

Figure 7: Aspen Plus SHIFTING hierarchy diagram to simulate the WGSR stage in the coal-SNG process.

A similar flow sheet presented in Figure 7 is used to simulate the water gas shift process of the dry feeding technology. The process takes into account that through stream S-9 flows around the 85%–95% of SYNGAS stream, while the remaining mass flow bypasses the WGSR reactor through stream S-10.

AGR Hierarchy

In this model, the acid gases recovery unit is modeled by means of a built-in SEP block (SEP-2). The clean-up process is simulated for the syngas by retiring all acid gases and other impurities formed upstream in the process, Figure 8. This stage is similar to both technologies simulated (dry and slurry).

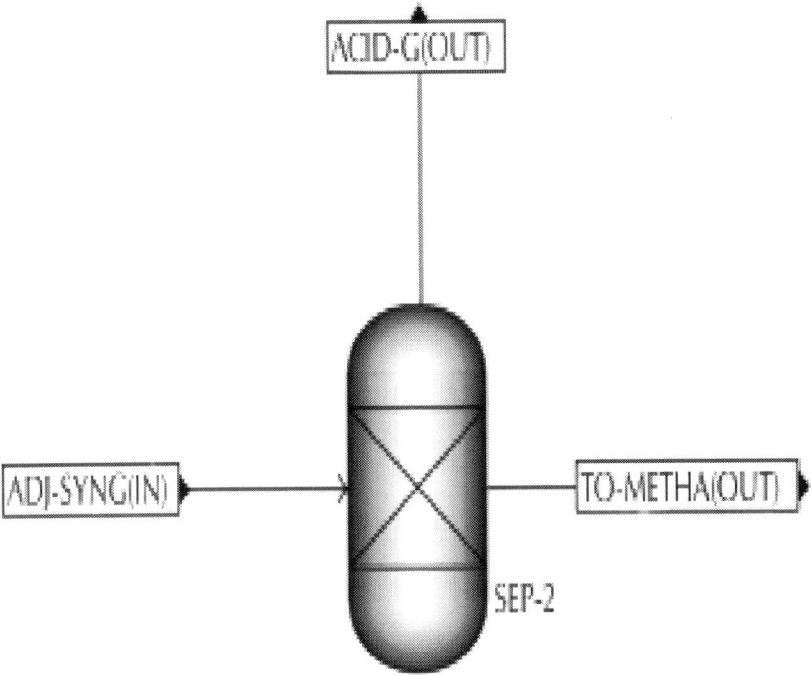

Figure 8: Aspen Plus AGR hierarchy diagram to simulate the acid gases clean-up in the coal-SNG process.

As shown in Figure 8, the ADJ-SYNG stream coming from SHIFTING hierarchy is separated into ACID-G stream (composed by H_2O, N_2, H_2S, NH_3, CO_2, and COS) and TO-METHA stream, which is composed mainly by CH_4, CO and H_2. The stream TO-METHA feeds the next hierarchy where methanation reaction is simulated.

METHANAT Hierarchy

The stream TO-METHA upcoming from the AGR hierarchy is driven to the METHANAT hierarchy, where promotion of methane formation is simulated. As shown in Figure 9, the stream TO-METHAN is fed into an equilibrium reactor (REQUIL-2), where CO and H_2 react to form CH_4 and H_2O (see (5)). Condensates are retired by stream S-14, while gaseous products are driven by stream S-12 into a heater (HEATER-3) where the gases are cooled and sent to SEP unit (SEP-3). In the SEP-3 separator, the split process between SNG and water is simulated. CO_2 and any other possible remaining impurities are retired by stream S-15. In this way, the stream SNG (Figure 9) contains the desired gaseous product with high methane concentration.

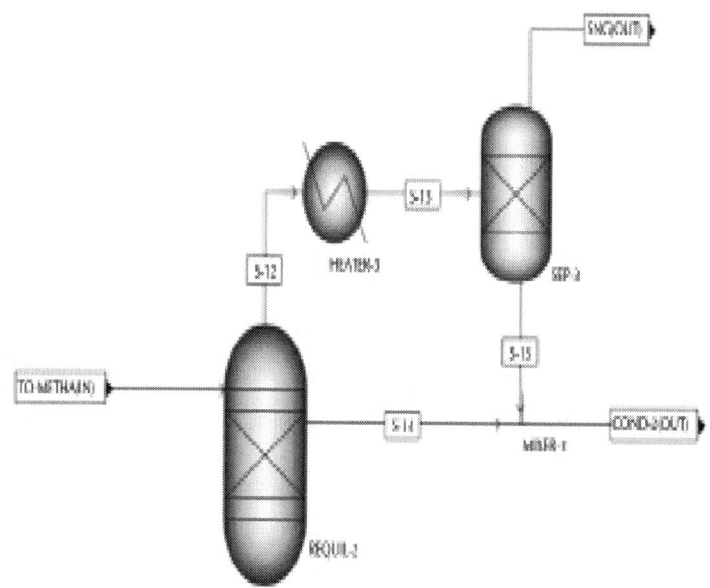

Figure 9: Aspen Plus METHANAT hierarchy diagram to simulate the methanation reaction in the cola-SNG process.

In this hierarchy, there are no model process differences between the slurry feeding and the dry feeding technology; thus, the dry feeding technology is simulated using the same diagram model (Figure 9).

Energy Parameters

Energy indicators were estimated to characterize the SNG production via coal gasification using Colombian coals as raw material. These parameters are useful to analyze the process in function of the feeding fuel system technology. The description of such energy indicators is given below.

- Equivalence ratio (ER), (7), accounts for the oxygen/coal ratio in the thermochemical process. ER > 1 represents a poor-fuel process while ER < 1 indicates a rich-fuel process or incomplete combustion. ER = 1 indicates stoichiometric combustion, where all the fuel in the gasifier is completely oxidized and transformed into H_2O and CO_2. Consider

$$ER = \frac{(\dot{m}_{O_2}/\dot{m}_{coal})}{ER_{stq}}, \quad (7)$$

Where $\dot{m}_{O2}$ and $\dot{m}_{coal}$ are the inlet mass flows of oxygen and coal, respectively. ER_{stq} is the stoichiometric equivalent ratio (see (8)) calculated with the theoretical combustion reaction (see (9)). Consider

$$ER_{stq} = \frac{\varphi \cdot M_{O_2}}{M_{coal}}, \quad (8)$$

$$C_nH_mO_pN_qS_r + \varphi O_2 \rightarrow aCO_2 + bH_2O + dN_2 + fSO_2, \quad (9)$$

Where e $C_nH_mO_pN_qS_r$ represents a recursive fuel substitution formula estimated in dry base and ash free; $\varphi = n + m/4 + r/2 - p/2$; MO_2 a and M_{coal} are the molecular weight of oxygen and coal, respectively.

- Coal Conversion efficiency (CCE, %), (10), is the ratio between the amount (mass units) of SNG at the exit of the process and the amount (mass units) of coal at the inlet of the process. Consider

$$CCE = \frac{\dot{m}_{SNG}}{\dot{m}_{coal}} \times 100 \quad (10)$$

with m_{SNG} the mass flow of substitute natural gas leaving the process.

- Cold gas efficiency (η_{cg}, %), or the energy efficiency of the gasification process, (11), is the ratio between the syngas energy in the stream leaving from the gasifier and the coal energy in the stream feeding into the process. Consider

$$\eta_{cg} = \frac{\dot{m}_{syngas} \cdot HHV_{syngas}}{\dot{m}_{coal} \cdot HHV_{coal}} \times 100, \quad (11)$$

where m_{syngas} is the mass flow of syngas leaving from the gasifier; HHV_{coal} is the higher heating value of coal as received that is including ash and moisture contents; HHV_{syngas} is the higher heating value of the syngas stream leaving the gasifier (wet basis), (12). Consider

$$HHV_{syngas} = \sum_{i=1}^{k} X_i \cdot HHV_i \quad (12)$$

with X = mass fraction (wet basis); i = each one of the gaseous species with energy contents considered in the syngas, that is, CO, H_2, CH_4, C_2H_4, C_6H_6, H_2S, and NH_3; HHV_i values are taken from the literature and are presented in Table 2 [50].

Table 2: High heating values of the gaseous species considered in the syngas leaving from the gasifier [50]

Gaseous specie	HHV (MJ/Nm3)	HHV (MJ/kg)
CO	12.622	10.1
H2	12.769	141.8
CH4	39.781	55.53
C2H4	63	50.2952
C6H6	142.893	41.8
H2S	25.105	16.488
NH3	13.072	22.428

- Process efficiency (η_{pro}, %), (13), is the energy efficiency of the conversion of the syngas to SNG, estimated as the ratio between energy of SNG on the stream leaving the process and syngas energy on the stream leaving the gasifier, where HHV_{SNG} is the higher heating value of the SNG. Consider

$$\eta_{pro} = \frac{\dot{m}_{SNG} \cdot HHV_{SNG}}{\dot{m}_{syngas} \cdot HHV_{syngas}} \times 100. \quad (13)$$

- Global efficiency (η_{global}, %, %), (14), is defined as the ratio between energy on the SNG stream leaving the process and energy in the coal stream feeding the process, that is, energy efficiency of the coal conversion to SNG. Consider

$$\eta_{global} = \frac{\dot{m}_{SNG} \cdot HHV_{SNG}}{\dot{m}_{coal} \cdot HHV_{coal}} \times 100 = \eta_{cg} \cdot \eta_{pro}. \quad (14)$$

- High heating value of the SNG (HHV_{SNG}), (15), accounts for the quality of the gas leaving the process. It is estimated from the molar fraction of gaseous products with energy content in the SNG stream. Consider

$$HHV_{SNG} = \sum_{j=1}^{n} Y_j \cdot HHV_j \quad (15)$$

with Y = molar fraction; j = each one of the gaseous species with energy content considered in the SNG, that is, CO, H_2, and CH_4.

- Wobbe Index, WI (MJ/Nm³), (16), accounts for the exchangeability of gases. According to the Wobbe Index, it is possible to know if the SNG quality is good enough to be transported by gas pipe lines, with density (ρ) at standard conditions (1 atm, 15°C). Consider

$$WI = \frac{HHV_{SNG}}{\sqrt{\rho_{SNG}/\rho_{air}}}. \tag{16}$$

- Power, P (kW), (17), is a parameter referring to SNG power, and it is estimated as the product between SNG mass flow rate produced (m_{SNG}, kg/s) and its higher heating value (HHV_{SNG}, kJ/kg). Consider

$$P = \dot{m}_{SNG} \cdot HHV_{SNG}. \tag{17}$$

Simulation Conditions

The mass flow rates presented in Table 3 are used to validate the model. The aim is to analyze the model accuracy in order to simulate the SNG production via coal gasification. The operation conditions (i.e., temperature and pressure) for reactors and other facilities (Table 4) were taken from the available technical report [51] or assumed in this work according to the literature or simulation results. The simulation conditions are defined to produce around 100 MMFCD SNG. Therefore, the pressure of the gasifier is higher than 40 bar due to the objective of this work being the study of a real and feasible process [52].

Table 3: Mass flow rates used in the model as inlet streams

Technology feeding	Raw material	Coal (ton/h)	Water (ton/h)	Oxygen (ton/h)	CO2 (ton/h)	Steam (ton/h)
Slurry	Bijao	293.6	126.3	173.9	—	86.7*
	Sanoha	219.5	107.1	138.02	—	105*
Dry	Bijao	237.2	—	175.2	143.5	189*
	Sanoha	220.4	—	188	138.2	192.7*

*Estimated in this work, using sensitivity analysis, to reach the ratio $H_2/CO = 3$.

Table 4: Reactors temperature and pressure conditions, inlet streams, and other model facilities described in the Aspen Plus models

Facility or inlet stream	Temperature (°C)	Pressure (bar)
Coal stream	27	1
Water stream	40	5
Oxygen stream	40	74.1
CO2 stream	100	83
Steam stream	250*	49*
Reactor RGIBB-1 (Figure 5)	Bijao coal: 1658** Sanoha coal: 1413**	50
Reactor RGIBB-2 (Figure 5)	Bijao coal: 940** Sanoha coal: 922**	50
Reactor RGIBB-1 (Figure 6)	Bijao coal: 1340** Sanoha coal: 1508**	50
Reactor REQUIL-1 (Figure 7)	250* (Bijao and Sanoha coals)	49*
Reactor REQUIL-2 (Figure 9)	350* (Bijao and Sanoha coals)	42*
Heater HEATER-1 (Figure 5)	202	Isobaric
Heater HEATER-1 (Figure 6)	167	Isobaric
Heater HEATER-2 (Figure 7)	40	Isobaric
Heater HEATER-3 (Figure 9)	54.7	Isobaric

*Adapted from [25]. **Simulation results.

The energy parameters, such as HHV_{syngas} and HHV_{SNG}, coal conversion efficiency, cold efficiency, process efficiency, global efficiency, Wobbe Index, and power (see (12)–(17)), were directly estimated within the Aspen Plus model by means of Fortran statements in calculator blocks. The equivalence ratio was estimated as an operation condition of the gasifier and thus the reaction temperatures.

RESULTS AND DISCUSSION

In this section, the simulation results for SNG synthesis with two Colombian coals (Bijao and Sanoha) are compared with available data from the technical report using slurry and dry feeding technologies [51]. Simulation results include the syngas and the SNG compositions, SNG flows (mass and volumetric flows), and energy parameters. These results are useful to compare the feeding technologies as well as fuel specifications (Colombian coal rank) in the SNG production via coal gasification.

Two global parameters to estimate the quality of simulation results and to compare with reference data are proposed in different works, the root mean square deviation (RMSD) or root mean square error (RMSE) and the relative error (RE) [53, 54]. These parameters are considered in this work to analyze the model accuracy against the reference data. The RMSD has the units of each parameter analyzed and the RE is expressed in percentage (%). The RE is presented as the average error between estimated and reported amount of each gaseous specie in the syngas, for both coal types and both feeding technologies. Consider

$$\text{RMSD} = \sqrt{\frac{\sum_{i=1}^{n}\left(X_{\text{reference},i} - X_{\text{model},i}\right)^2}{n}},$$

$$\text{RE} = \left|\frac{X_{\text{reference},i} - X_{\text{model},i}}{X_{\text{reference},i}}\right| \cdot 100. \qquad (18)$$

Syngas and SNG Composition

The simulated syngas composition for both feeding technologies is validated by comparison with technical report data and it is shown in Table 5. To estimate the model behavior, the RMSD and the RE are presented in the same table. Both comparative parameters have been estimated for each specie in the gas (syngas and SNG) including the coal type and feeding technology in the same calculus.

Table 5: Validation of syngas composition (vol% dry base) for two different Colombian coals under slurry and dry feeding technologies

Syngas (%vol d.b.)	Slurry feeding				Dry feeding				RMSD (units)	$\overline{RE}^*$ (%)
	Bijao		Sanoha		Bijao		Sanoha			
	[51]	Model	[51]	Model	[51]	Model	[51]	Model		
N_2	0.61	0.35	0.6	0.3	1.27	0.628	1.26	0.61	3.6	48.7
H_2	37.10	35.39	30.9	38.52	26.06	20.58	26.25	20.19	5.3	18.3
CO	37.88	37.07	44.9	42.27	64.49	69.45	65.29	72.41	3.3	6.6
CO_2	20.64	22.08	14.0	13.98	7.6	8.80	6.61	6.26	1.5	7.1
CH_4	3.24	3.89	9.0	3.74	0.01	0.02	0.01	0.01	2.7	44.6
H_2S	0.44	0.44	0.5	0.42	0.42	0.41	0.42	0.41	0.8	5.2
NH_3	0.01	0.7	0.0	0.71	0.03	0.002	0.03	0.00	5.0	2364.4
COS	0.02	0.02	0.0	0.02	0.03	0.06	0.03	0.07	5.8	77.8
Ar	0.04	0.05	0.0	0.04	0.05	0.04	0.06	0.05	2.0	20.6
$HHV_{w.b.}$ (MJ/Nm3)	8.49	7.44	11.77	9.06	9.28	10.66	11.47	11.1	1.6	13.4

*Calculated as the average of the four pairs simulation-reported data described in each entire row.

The coherent behavior of the model to predict the species with higher concentration in the syngas stream is observed. The RE of H_2, CO, and CO_2 between simulated and reference data varies from 6.6% to 18%. This is confirmed in Table 5 by the lower RMSD of the volumetric concentration of these gaseous species. For example, the RMSD of H_2 indicates that the model prediction varies ±5.3%vol; this value is lower compared with the H_2 concentration produced in the entrained flow gasifiers simulated. The RMSD of CO and CO_2 behaves in a similar way. Nevertheless, the model cannot predict with accuracy the species with lower concentration (traces of N_2, CH_4, NH_3, and COS) in the syngas due to their higher RE. This is also corroborated with the RMSD showed in Table 5. The model variability is higher against the traces concentrations reached in the gasification technologies simulated.

It is important to highlight that the temperature operation of the gasifier was not available in the reference work and thus could be an important deviation source between the reference and simulated data.

Regarding the HHV_{syngas} (wet basis), a fundamental energy parameter to characterize the gasification process, it is highlighted that the model is able to predict coherently this factor, because its RE is 13% and its RMSD varies in ±1.6 MJ/Nm³. The higher heating value is around 10 MJ/Nm³ for both raw materials and both technologies simulated. These results are quite similar to other coal gasification reports that use the chemical equilibrium approach for their simulations [24].

Despite the low accuracy of the model to estimate traces but considering the HHV_{syngas} as global parameter of the process, it is possible to state that the gasification model behaved coherently under different coal types and feedstock feeding technologies.

Table 6 compares the composition of SNG between simulated conditions in this work and reported data [51]. Table 7 shows the mass and volumetric flows of SNG produced as well as the HHV_{SNG} (see (15)), Wobbe Index (see (16)), and power (see (17)).

Table 6: Comparison between SNG compositions simulated and reported data under slurry and dry feeding technologies

SNG (% vol.)	Slurry feeding				Dry feeding				RMSD (units)
	Bijao		Sanoha		Bijao		Sanoha		
	[51]	Model	[51]	Model	[51]	Model	[51]	Model	
H2	0.33	1.17	0.32	1.07	0.31	1.11	0.31	1.22	0.8
CH4	97.83	95.98	98.2	96.55	94.11	96.02	94.18	96.05	1.8
CO2	0.07	0.00	0.07	0.00	0.06	0	0.06	0.00	0.1
N2	1.55	2.68	1.23	2.25	5.3	2.69	5.19	2.54	2.0
Ar	0.23	0.17	0.17	0.13	0.21	0.18	0.26	0.19	0.1

Table 7: Model validation with regard to SNG production and quality as gaseous fuel

Energy parameter	Slurry feeding				Dry feeding				RMSD (units)	$\overline{RE}$ (%)
	Bijao		Sanoha		Bijao		Sanoha			
	[51]	Model	[51]	Model	[51]	Model	[51]	Model		
HHVSNG (MJ/Nm3)	39	38.33	39.1	38.55	37.5	38.34	37.5	38.37	0.74	1.9
Mass flow (kg/h)	78400	81006	82700	86063	79300	80504.8	82000	80332	2362.98	2.7
Volumetric flow (MMFCD)	96.81	101.18	102.51	107.49	95.3	100.55	98.56	100.34	4.32	4.2
WI (MJ/m3)	51.6	51.68	51.9	51.98	49	51.70	49	51.73	1.92	2.8
Power (MW)	1147	1250	1221	1328	1107	1242	1146	1239	110.6	9.5

*Calculated as the average of the four pairs simulation-reported data described in each entire row.

concentrations in the simulated SNG are comparable for both technologies (Table 6) independent of coal type. The average RMSD of SNG composition is ±1.8%vol and the average relative error between reported and simulated data is lower than 2%. However, the yield of CH_4 with slurry feeding technology tends to be slightly higher due to the two gasification stages. Methane composition in the SNG simulated for both technologies is closer to Colombian natural gas composition from the Guajira region (around 97%vol of CH_4).

On the other hand, in all simulated cases the model is not able to predict the traces concentration of H_2, N_2, and CO_2 in the SNG stream. The RE of simulated data for these species compared with the reference data is higher than 65%. This can be seen in Table 6, where the RMSD of the traces is higher compared with the concentration produced in the gasification process. However, it is important to highlight the relative small magnitude order for H_2, Ar, N_2, and CO_2 concentration levels (lower than 2%vol) against to the CH_4 concentration in the SNG (between 94 and 98%vol). Therefore, the traces concentration has a negligible effect on the energy parameters estimated in Section 4.2.

SNG Production and Quality as Gaseous Fuel

Despite the model low accuracy to predict the traces in the SNG stream, in Table 7, it can be observed that HHV_{SNG} calculated by the simulations is almost equal for both Sanoja and Bijao Colombian coals (independent of the feeding technology), and in both cases HHV_{SNG} is quite similar to the reported data in the literature. The average relative error of HHV_{SNG} is lower than 2%; this can be explained due to similar CH_4 contents (Table 6) obtained from methanation stage. Thereby, feeding technology would not affect HHV_{SNG}. According to HHV_{SNG} and WI_{SNG} presented in Table 7, the simulated SNG can be classified as a high quality natural gas (NG) and it can be transported by gas pipe lines (for both coals and both technologies used) [22]. The high quality classification is due to the fact blank that HHV_{SNG} and WI_{SNG} parameters are between the ranges defined by Colombian regulatory norms. HHV_{NG} must be between 35.4 and 42.8 MJ/m³ and WI 47.7–52.7 MJ/m³ [55]. Therefore, the SNG produced and distributed can be used in transport, power, and heat generation in industrial and domestic systems [21]. Thereby, the SNG can contribute to meet the increasingly high demand of the gaseous fuel in Colombia.

It is evidenced in Table 7 that for both feeding technologies the simulation of SNG production from Colombian coal gasification does not show significant differences (in mass flows or volumetric flows) when compared with reference report [51]. This is due to the coherent RE that varies between 2.7% and 4.2%. Therefore, the models developed in this work can be considered as computational tools to study the coal to SNG process. The differences between simulated and reference data can be attributed to the unavailability of STEAM flow rates in the reported data (Table 4) and its incidence in global mass balances.

Regarding the power of the SNG (Table 7), the values calculated are between the expected error boundaries due to mass flow and HHV differences among simulations and reported data. The power is overestimated by 9.5% with both coals without relying on the technology. The higher potential energy obtained for Sanoha coal is due to its better quality as fuel (Table 1).

Mass and Energy Efficiencies

The carbon conversion efficiency (CCE) is presented in Figure 10. It is observed that Sanoha coal tends to give higher coal conversion, especially with the slurry feeding technology. This could be explained by the higher volatile matter contents of this coal (Table 1); and because of the slurry feeding technology, the first one of two gasification stages operates at higher temperatures than the gasifier dry feeding technology (Table 4).

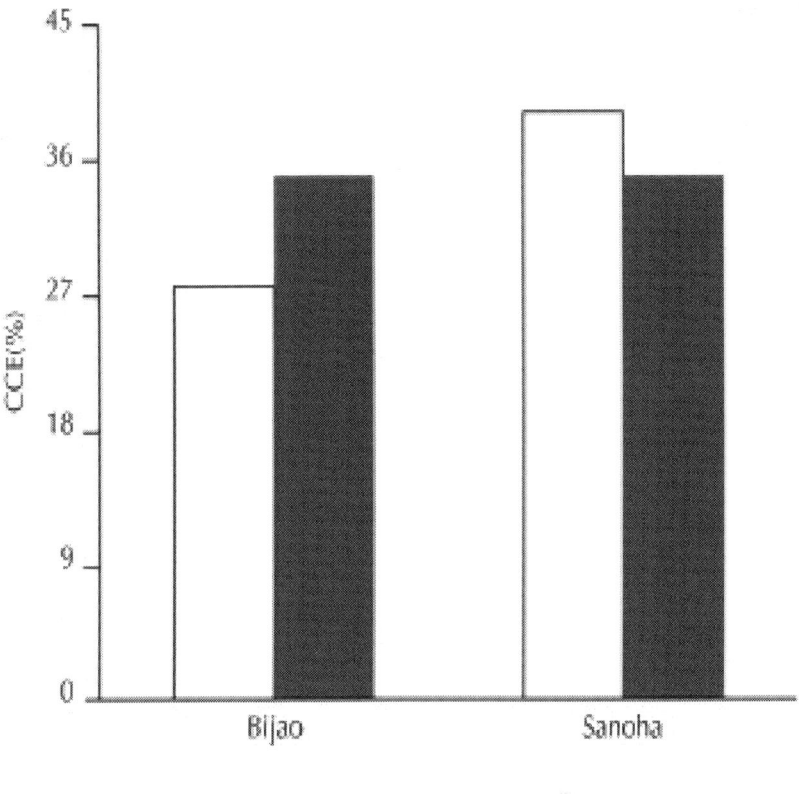

(a)

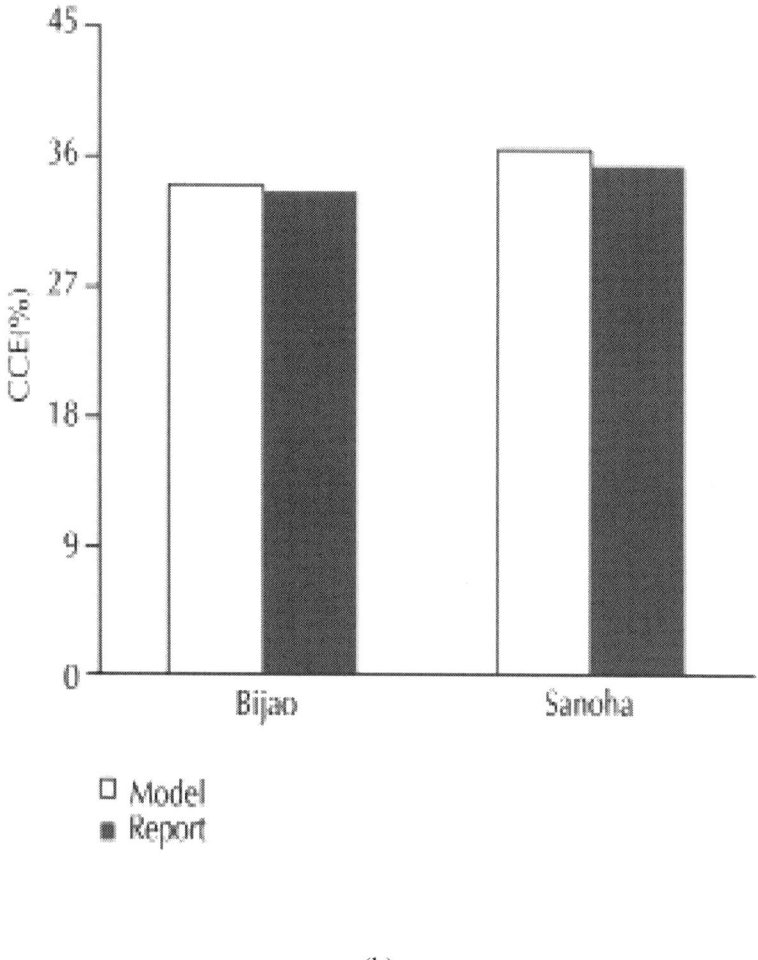

(b)

Figure 10: Carbon conversion efficiency (kg CH_4/kg coal). Comparison for simulated and reported data for two Colombian coals; (a) slurry feeding technology and (b) dry feeding technology.

The difference between simulated CCE and technical report data is about 3.5%. Moreover, the simulated data agree with biomass gasification yields reported in the literature, around 26% kg CH_4/kg wood [25]. Differences of CCE around 3%-4% between coal and wood as raw material can be explained by the coal higher fixed carbon with regards to wood or biomass.

It is observed from Figure 11 that cold gas efficiency is slightly higher for dry feeding technology, as expected according to the literature [12]. This trend is due to the slurry feeding technology requiring more energy for the autothermal process to vaporize water in the gasifier. Nevertheless, relative differences between reported and simulated data are almost meaningless, independent of the coal or the simulated technology.

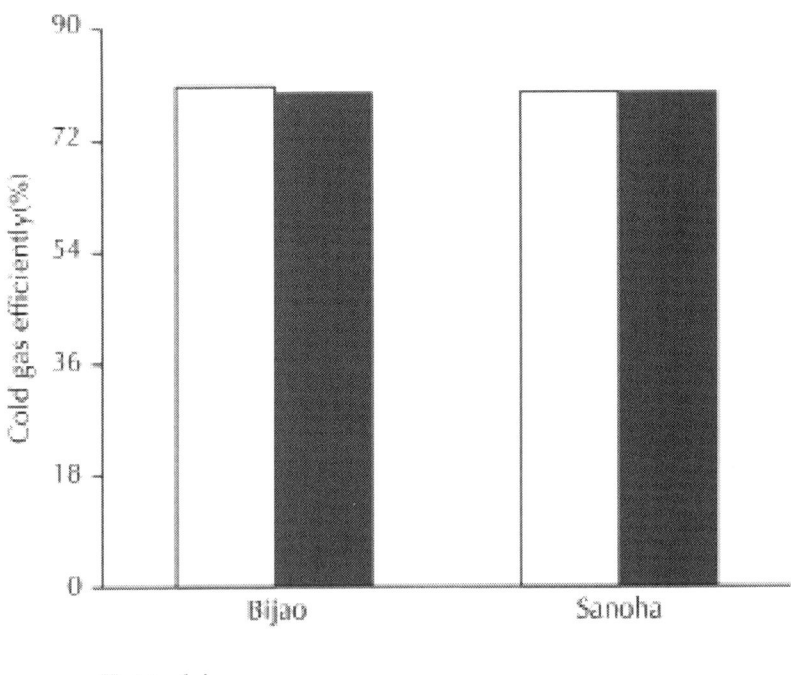

(a)

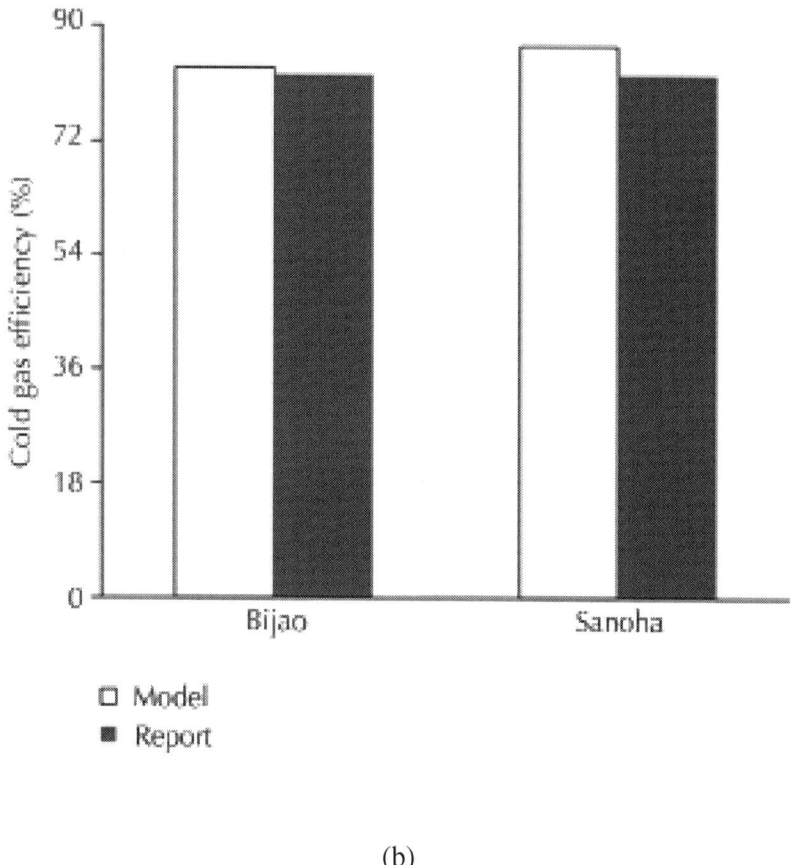

(b)

Figure 11: Cold gas efficiency (%). Comparison between simulated and reported data for two Colombian coals; (a) slurry feeding technology and (b) dry feeding technology.

Figure 12 shows that process efficiency is higher for the slurry feeding technology when compared with dry feeding technology, independent of the coal. This is explained because the higher methane content in syngas from slurry technology produces a higher syngas quality to be fed into the catalytic methane reactor. The same trend is observed in the comparison of global efficiencies, Figure 13.

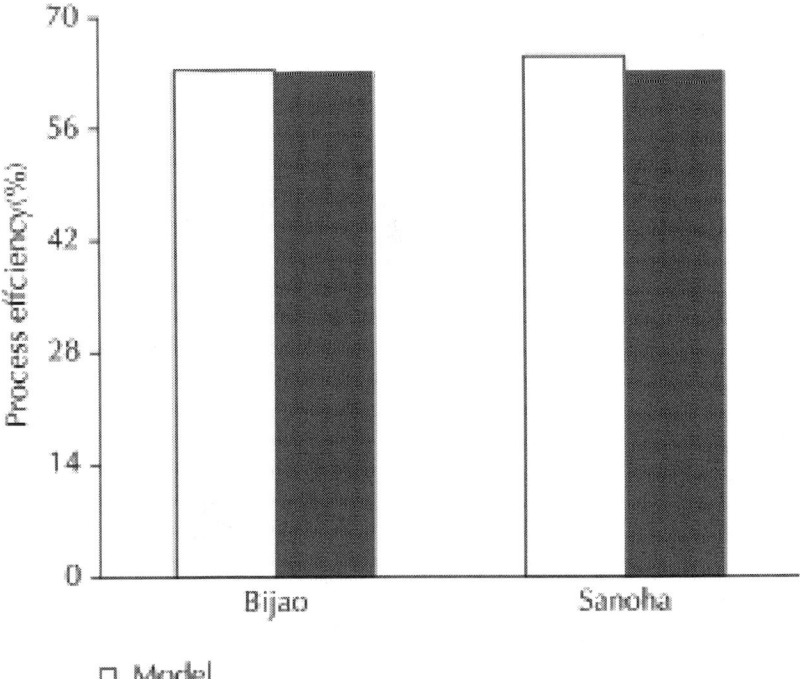

(a)

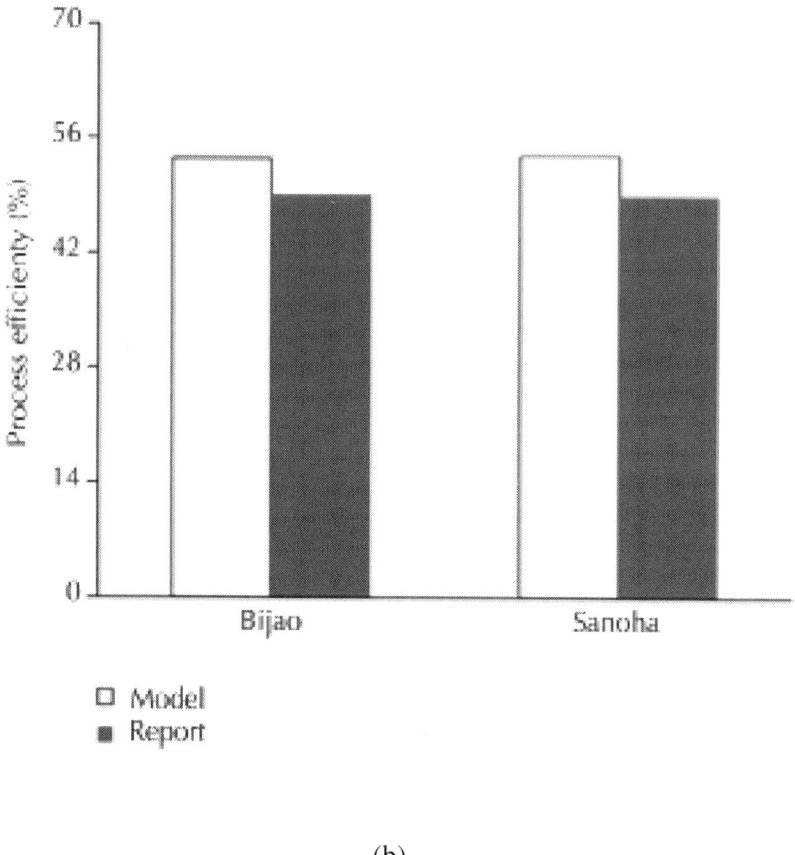

(b)

Figure 12: Process efficiency (%). Comparison between simulated and reported data for two Colombian coals; (a) slurry feeding technology and (b) dry feeding technology.

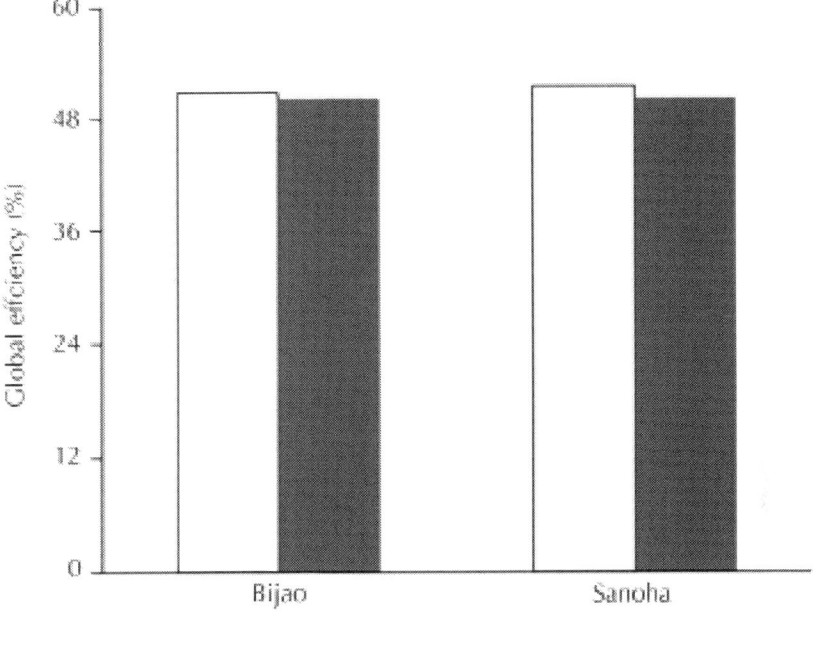

(a)

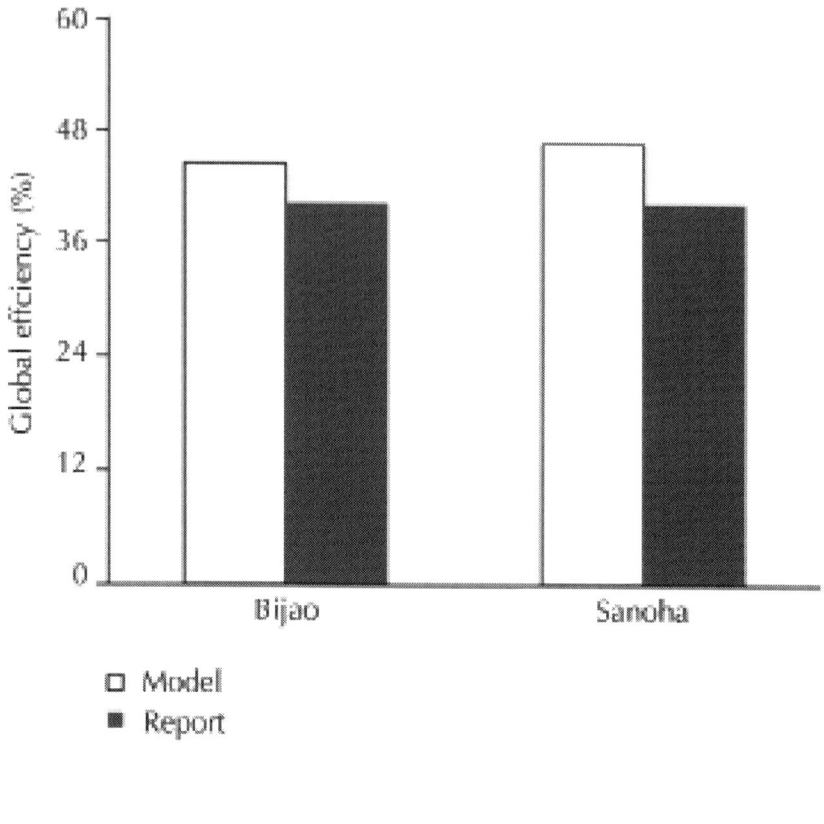

(b)

Figure 13: Global efficiency (%). Comparison between simulated and reported data for two Colombian coals; (a) slurry feeding technology and (b) dry feeding technology.

From Figures 10–13, it can be observed that even when efficiencies estimated by the simulations with dry technology are slightly lower than those obtained with slurry feeding technology, the average margin of error between the model and technical report is less than 5%. It is clear that the process efficiency (Figure 12) is 3% higher for Sanoha coal with regard to Bijao coal. This is due to the Sanoha coal having a higher quality rank than Bijao coal (Table 1).

The process and global efficiencies are higher for slurry feeding technology (Figures 12 and 13). This is due to the higher hydrogen concentration in the syngas produced with the slurry feed technology against dry feed technology [21].

The slurry feeding technology tends to be more efficient than dry feeding technology in the process of coal to SNG simulated in this work. Differences in these parameters were not found in function of the coal rank. The cold gas efficiency is similar to all simulated cases with a slightly improvement for dry feeding technology. The gasification efficiency is better for Sanoha coal due to its fixed carbon content and higher HHV_{coal} (Tables 1 and 5).

The RMSD was estimated accounting all data presented. Low differences between simulated and reported data, for both coals using the two different feeding technologies, were found. These values indicate a good agreement between simulated data with the models described in this study and reported data available from the literature. Therefore, the model developed in this paper constitutes a valuable and versatile tool to study the thermodynamic performance of the solid fuels to SNG process via gasification.

CONCLUSIONS

A model to simulate the SNG production via coal gasification process has been developed using Aspen Plus software. The model considers two different typical technologies to supply the fuel into the gasification process. The feeding systems are coal-water slurry and dry coal. The model can simulate different solid fuel types. Therefore, the effect of using two different Colombian coals, Bijao and Sanoha, is considered in this work.

A global comparison between reported and simulated data presents an average relative error lower than 13%. Therefore, the developed model is able to predict the composition and heating values of syngas and SNG and the SNG quality and energy parameters of the process and the trends. The computational model presented in this work and developed in Aspen Plus v7.3 software can be used for gaining a fundamental understanding of the engineering and optimization of the process, even when scaled up.

According to the simulation process, it was found that coal rank does not significantly affect energy indicators such as cold gas, process, and global efficiencies. However, feeding technology clearly has an effect on these energy parameters. The process and global efficiencies are higher for slurry feeding technology, while cold gas efficiency was higher for dry feeding technology; these results agree with the literature. According to HHV_{SNG} and WI, the simulated SNG from both coals and both technologies can be classified as a high quality NG in Colombia. Therefore, the gaseous fuel can be transported by gas pipe lines and the SNG can contribute to meet the increasingly high demand of the gaseous fuel in Colombia.

Since the proposed model can be used to analyze various types of entrained flow reactors with different operating conditions, it can be considered a versatile and useful computational tool to optimize the coal to SNG process. In a future work, a sensitivity analysis of the effect of Colombian coal rank (subbituminous to semi-anthracite) will be conducted.

ACKNOWLEDGMENTS

The authors would like to thank the energy Colombian company Celsia S.A. ESP for the financial support of the project: developing and validation of a computational model to simulate the production of synthetic natural gas by means of coal gasification with Aspen Plus: effect of the Colombian coals rank (in Spanish)—code: PI12-1-05.

REFERENCES

1. M. Höök and K. Aleklett, "A review on coal-to-liquid fuels and its coal consumption," International Journal of Energy Research, vol. 34, no. 10, pp. 848–864, 2010
2. J. Lu, L. Yu, X. Zhang, S. Zhang, and W. Dai, "Hydrogen production from a fluidized-bed coal gasifier with in situ fixation of CO_2. Part I. Numerical model," Chemical Engineering and Technology, vol. 31, no. 2, pp. 197–207, 2008.
3. P. J. Robinson and W. L. Luyben, "Simple dynamic gasifier model that runs in aspen dynamics," Industrial and Engineering Chemistry Research, vol. 47, no. 20, pp. 7784–7792, 2008.

4. Z. Yuehong, W. Hao, and X. Zhihong, "Conceptual design and simulation study of a co-gasification technology," Energy Conversion and Management, vol. 47, no. 11-12, pp. 1416–1428, 2006.
5. NETL Gasifipedia, Coal Power Gasification, 2013, http://www.netl.doe.gov/technologies/coalpower/gasification/gasifipedia/.
6. J. Lee, S. Park, H. Seo et al., "Effects of burner type on a bench-scale entrained flow gasifier and conceptual modeling of the system with Aspen Plus," Korean Journal of Chemical Engineering, vol. 29, no. 5, pp. 574–582, 2012
7. IEA Clean Coal Centre, Future Development of IGCC, 2008, http://www.iea-coal.org.uk/documents/82119/7089/Future-developments-in-IGCC.
8. Unidad de Planeación Minero Energética (UPME) and Ministerio de Minas y Energía (Colombia), "Plan de Abastecimiento para el Suministro y Transporte de Gas Natural," 2010, http://www.upme.gov.co/Docs/Plan_Abast_Gas_Natural/PLAN_ABASTECIMIENTO_GAS%20NATURAL_2009.pdf.
9. A. Martínez, "La actualidad del gas natural en Colombia," Revista Petrotecnia, 2010, http://www.petrotecnia.com.ar.
10. British Petroleum, BP Statistical Review of World Energy, 2011, http://www.bp.com/content/dam/bp-country/de_de/PDFs/brochures/statistical_review_of_world_energy_full_report_2011.pdf.
11. M. Gräbner and B. Meyer, "Performance and exergy analysis of the current developments in coal gasification technology," Fuel, vol. 116, pp. 910–920, 2014.
12. C. Kunze and H. Spliethoff, "Modelling, comparison and operation experiences of entrained flow gasifier," Energy Conversion and Management, vol. 52, no. 5, pp. 2135–2141, 2011.
13. N. Seifkar, W. Davey, and J. Sarvinis, "Comparison of several coal gasification processes," in Proceedings of the 8th World Congress of Chemical Engineering (WCCE '09), August 2009.
14. O. Maurstad, H. Herzog, O. Bolland, and J. Beér, "Impact of coal quality and gasifier technology on IGCC performance," Norwegian Research Council in the KLIMATEK program, 2013, http://sequestration.mit.edu/pdf/GHGT8_Maurstad.pdf.

15. G. W. Yu, Y. M. Wang, and Y. Y. Xu, "Modeling analysis of shell, Texaco gasification technology's effects on water gas shift for Fischer-Tropsch process," Advanced Materials Research, vol. 608-609, pp. 1446–1453, 2012.

16. M. Prins, R. van den Berg, E. van Holthoon, E. van Dorst, and F. Geuzebroek, "Technological developments in IGCC for carbon capture," Chemical Engineering and Technology, vol. 35, no. 3, pp. 413–419, 2012

17. M. J. Bockelie, M. K. Denison, Z. Chen, C. L. Senior, and A. F. Sarofim, "Using Models to Select Operating Conditions for Gasifiers," Reaction Engineering International, http://energy.reaction-eng.com/downloads/REI_processmodel.pdf.

18. S. Armin, Simulation of Coal Gasification Process Inside a Two-Stage Gasifier, Department of Mechanical Engineering, University of New Orleans, New Orleans, La, USA, 2004.

19. A. Silaen and T. Wang, "Investigation of the coal gasification process under various operating conditions inside a two-stage entrained flow gasifier," Journal of Thermal Science and Engineering Applications, vol. 4, no. 2, Article ID 021006, 2012.

20. C. R. Vitasari, M. Jurascik, and K. J. Ptasinski, "Exergy analysis of biomass-to-synthetic natural gas (SNG) process via indirect gasification of various biomass feedstock," Energy, vol. 36, no. 6, pp. 3825–3837, 2011.

21. A. Tremel, M. Gaderer, and H. Spliethoff, "Small-scale production of synthetic natural gas by allothermal biomass gasification," International Journal of Energy Research, vol. 37, no. 11, pp. 1318–1330, 2013.

22. S. Heyne, H. Thunman, and S. Harvey, "Extending existing combined heat and power plants for synthetic natural gas production," International Journal of Energy Research, vol. 36, no. 5, pp. 670–681, 2012

23. B. James, Cost and Performance Baseline for Fossil Energy Plants, vol. 1 of Bituminous Coal and Natural Gas to Electricity, DOE-National Energy Technology Laboratory, 2010, http://www.netl.doe.gov/research/energy-analysis/energy-baseline-studies.

24. F. Trippe, M. Fröhling, F. Schultmann, R. Stahl, and E. Henrich, "Techno-economic assessment of gasification as a process step

within biomass-to-liquid (BtL) fuel and chemicals production," Fuel Processing Technology, vol. 92, no. 11, pp. 2169–2184, 2011.

25. K. J. Ptasinski, A. Sues, and M. Jurascik, "Biowates to biofuels: routes via gasification," inBiomass Gasification: Chemistry, Processes and Applications, J. P. Badea and A. Levi, Eds., pp. 86–197, Nova Science, New York, NY, USA, 2009.

26. O. Maurstad, "An overview of coal based integrated gasification combined cycle (IGCC) technology," Publication No. LFEE 2005-002 WP, Massachusetts Institute of Technology, 2005, http://sequestration.mit.edu/pdf/LFEE_2005-002_WP.pdf.

27. Aspen Tech Web Site, http://www.aspentech.com/products/aspen-plus.aspx.

28. P. L. Douglas and B. E. Young, "Modelling and simulation of an AFBC steam heating plant using ASPEN/SP," Fuel, vol. 70, no. 2, pp. 145–154, 1991.

29. C. E. Backham and P. L. Douglas, "Simulation of a coal hydrogasification process with integrated CO_2 capture," Combust Canada, 3A, 2003.

30. W. Doherty, A. Reynolds, and D. Kennedy, "The effect of air preheating in a biomass CFB gasifier using ASPEN Plus simulation," Biomass and Bioenergy, vol. 33, no. 9, pp. 1158–1167, 2009.

31. R. Nayak and R. Mewada, "Simulation of coal gasification process using ASPEN PLUS," in Proceedings of the International Conference On Current Trends In Technology: Nuicone, New Delhi, India, 2011.

32. W. Doherty, A. Reynolds, and D. Kennedy, "Simulation of a circulating fluidised bed biomass gasifier using ASPEN plus—a performance analysis," in Proceedings of the 21st International Conference on Efficiency, Cost, Optimization, Simulation and Environmental Impact of Energy Systems (ECOS '08), Kraków, Poland, 2008.

33. A. Ong'iro, V. Ismet Ugursal, A. M. Al Taweel, and G. Lajeunesse, "Thermodynamic simulation and evaluation of a steam CHP plant using ASPEN plus," Applied Thermal Engineering, vol. 16, no. 3, pp. 263–271, 1996.

34. M. Gazzani, G. Manzolini, E. MacChi, and A. F. Ghoniem, "Reduced order modeling of the Shell-Prenflo entrained flow gasifier," Fuel, vol. 104, pp. 822–837, 2013.
35. S. Karellas, K. D. Panopoulos, G. Panousis, A. Rigas, J. Karl, and E. Kakaras, "An evaluation of Substitute natural gas production from different coal gasification processes based on modeling," Energy, vol. 45, no. 1, pp. 183–194, 2012
36. Q. Yi, J. Feng, and W. Y. Li, "Optimization and efficiency analysis of polygeneration system with coke-oven gas and coal gasified gas by Aspen Plus," Fuel, vol. 96, pp. 131–140, 2012
37. A. O. Ong'iro, V. I. Ugursal, A. M. Al Taweel, and D. K. Blamire, "Simulation of combined cycle power plants using the ASPEN PLUS shell," Heat Recovery Systems and CHP, vol. 15, no. 2, pp. 105–113, 1995.
38. S. Cimini, M. Prisciandaro, and D. Barba, "Simulation of a waste incineration process with flue-gas cleaning and heat recovery sections using Aspen Plus," Waste Management, vol. 25, no. 2, pp. 171–175, 2005
39. S. Chern, L. T. Fan, and W. P. Walawender, "Analytical calculation of equilibrium gas composition in a C-H-O-inert system," AIChE Journal, vol. 35, no. 4, pp. 673–675, 1989.
40. K. G. Mansaray, A. E. Ghaly, A. M. Al-Taweel, F. Hamdullahpur, and V. I. Ugursal, "Mathematical modeling of a fluidized bed rice husk gasifier: part I—model development," Energy Sources, vol. 22, no. 1, pp. 83–98, 2000.
41. J. Sadhukhan, K. S. Ng, N. Shah, and H. J. Simons, "Heat integration strategy for economic production of combined heat and power from biomass waste," Energy and Fuels, vol. 23, no. 10, pp. 5106–5120, 2009.
42. L. Shen, Y. Gao, and J. Xiao, "Simulation of hydrogen production from biomass gasification in interconnected fluidized beds," Biomass & Bioenergy, vol. 32, no. 2, pp. 120–127, 2008.
43. P. Ji, W. Feng, and B. Chen, "Production of ultrapure hydrogen from biomass gasification with air," Chemical Engineering Science, vol. 64, no. 3, pp. 582–592, 2009.
44. J. Sadhukhan, Y. Zhao, M. Leach, N. P. Brandon, and N. Shah, "Energy Integration and analysis of solid oxide fuel cell based

microcombined heat and power systems and other renewable systems using biomass waste derived syngas," Industrial and Engineering Chemistry Research, vol. 49, no. 22, pp. 11506–11516, 2010.

45. F. Emun, M. Gadalla, T. Majozi, and D. Boer, "Integrated gasification combined cycle (IGCC) process simulation and optimization," Computers and Chemical Engineering, vol. 34, no. 3, pp. 331–338, 2010.

46. C. Kunze and H. Spliethoff, "Modelling of an IGCC plant with carbon capture for 2020,"Fuel Processing Technology, vol. 91, no. 8, pp. 934–941, 2010.

47. N. Ramzan, A. Ashraf, S. Naveed, and A. Malik, "Simulation of hybrid biomass gasification using Aspen plus: a comparative performance analysis for food, municipal solid and poultry waste," Biomass and Bioenergy, vol. 35, no. 9, pp. 3962–3969, 2011.

48. P. Se-Ik, L. Joon-Won, and S. Hea-Kyung, "Effects of different coal type on gasification characteristics," Transactions of the Korean Hydrogen and New Energy Society, vol. 21, pp. 470–477, 2010.

49. S. Park, J. Lee, H. Seo, G. Kim, and K. Kim, "Experimental investigations of the effect of coal type and coal burner with different oxygen supply angles on gasification characteristics," Fuel Processing Technology, vol. 92, no. 7, pp. 1374–1379, 2011.

50. L. Waldheim and T. Nilsson, "Heating value of gases from biomass gasification," Report Prepared for IEA Bioenergy Agreement Subcommittee on Thermal Gasification of Biomass, Task 20—Thermal Gasification of Biomass, 2001.

51. "Coal to SNG Feasibility Study. For the energy Colombian Company Celsia S.A. ESP," Tech. Rep. 60T03800, Jacobs Engineering Group, 2013.

52. X. Hao, G. Dong, Y. Yang, Y. Xu, and Y. Li, "Coal to Liquid (CTL): commercialization prospects in China," Chemical Engineering & Technology, vol. 30, no. 9, pp. 1157–1165, 2007

53. S. Jarungthammachote and A. Dutta, "Thermodynamic equilibrium model and second law analysis of a downdraft waste gasifier," Energy, vol. 32, no. 9, pp. 1660–1669, 2007.

54. M. Vaezi, M. Passandideh-Fard, M. Moghiman, and M. Charmchi, "On a methodology for selecting biomass materials for gasification purposes," Fuel Processing Technology, vol. 98, pp. 74–81, 2012.
55. CREG, Comisión Reguladora de Energía y Gas de Colombia, Resolución 071, Ministerio de Minas y Energía, 1999, http://www.creg.gov.co/html/i_portals/index.php.

Chapter 4

Numerical Simulation of an Industrial Absorber for Dehydration of Natural Gas Using Triethylene Glycol

Kenneth Kekpugile Dagde and Jackson Gunorubon Akpa

Department of Chemical/Petrochemical Engineering, Rivers State University of Science & Technology, Nkpolu, Port Harcourt, Rivers State, Nigeria

ABSTRACT

Models of an absorber for dehydration of natural gas using triethylene glycol are presented. The models were developed by applying the law of conservation of mass and energy to predict the variation of water content of gas and the temperature of the gas and liquid with time along the packing height. The models were integrated numerically using the finite divided difference scheme and incorporated into the MATLAB code. The results obtained agreed reasonably well with industrial plant data obtained from an SPDC TEG unit in Niger-Delta, Nigeria. Model prediction showed a percentage deviation of 8.65% for gas water content and 3.41% and 9.18% for exit temperature of gas and liquid, respectively.

INTRODUCTION

Natural gas needs to be dried before pipeline transport, because the water molecules present in the gas in both vapour and liquid state form hydrates which cause flow restrictions and pressure drops and lower the heating value of gas and corrode pipelines and other equipment. Other problems associated with the presence of water molecules are foaming, degradation, puking, corrosion, low pH, oxidation, thermal decomposition, inadequate absorber design for flow conditions, and salt contamination. Extensive literature is available on common gas dehydration systems including solid and liquid desiccant and refrigeration-based systems [1, 2]. There are several methods of dehydrating natural gas. The most common of these are liquid desiccant (glycol) dehydration and solid desiccant dehydration [3, 4]. Among these gas dehydration processes, absorption is the most common technique, where the water vapor in the gas stream becomes absorbed in a liquid solvent stream. Glycols are the most widely used absorption liquids as they approximate the properties that meet commercial application criteria [5, 6]. Several glycols have been found suitable for commercial application. Triethylene glycol (TEG) is by far the most common liquid desiccant used in natural gas dehydration as it exhibits most of the desirable criteria of commercial suitability [2]. The glycol absorber (contactor) contains trays that provide an adequate intimate contact area between the gas and the glycol. One other option to the tray TEG contactor is the use of structured packing. Structured packing was developed as an alternative to random packing to improve mass transfer control by use of a fixed orientation of the transfer surface. The combination of high gas capacity and reduced height of an equilibrium stage, compared with tray contactors, makes the application of structured packing desirable for both new contactor designs and existing tray contactor capacity upgrades. Hence, the structured packing may offer potential cost savings over trays [1].

Optimization of glycol dehydration unit of a natural gas plant is generally aimed at developing a suitable mathematical model which, when tested with plant data, will aid in deciding the best operating conditions required to reduce natural gas water content to the standard pipeline specification of less than 7 lb H_2O/MMSCF of gas [7, 8]. Triethylene glycol (TEG) would be used as the absorbent for this process and would be regenerated in a glycol dehydration unit to 99% purity. This, however, is not the case in most of these units. Jaćimović et al. [9] simulated a reactive absorption system for the absorption of CO_2 in a packed column using methyl diethanolamine (MDEA) as the solvent. Steady state conditions and plug flow were assumed for the gas phase, leading to a set of ordinary differential equations. In Richardson et al. [10], a mathematical model for the wet scrubbing of CO_2 using chilled ammonia was studied. Diffusion and conduction terms were included in the development of the unsteady state models. These models predict the variation of the concentration of the reactants and products with time across the packed height, as well as the variation of the temperature of the system with time across the packed height. The partial differential equations developed were solved using the numerical technique of MATLAB by applying the Robin, Neumann, and Dirichlet boundary conditions (BC) [11]. A similar study on CO_2 absorption was carried out by Ahmed et al. [12] using a highly concentrated monoethanolamine (MEA). Most studies on gas dehydration using TEG were simulated using special packages like HYSYS used by Peyghambarzadeh and Jafarpour [13] and the parameters used in their models cannot be easily obtained without extensive experimental studies; thus the model cannot be adapted for simulation of industrial absorber unit. In this paper, models for a functional industrial absorber are presented. The results from the models are compared with data obtained from functional full-scale industrial absorber plant.

MODEL DEVELOPMENT

The most common method for dehydration in the natural gas industry is the use of a liquid desiccant contactor (absorber) process. In this process, the wet gas is contacted with lean solvent (triethylene glycol) as the absorbent. The water in the gas is absorbed in the lean solvent, producing a rich solvent stream and a dry gas. The dehydrated gas leaves at the top of the column while the glycol leaves at the bottom. Figure 1 depicts the hypothetical representation of the dehydrator.

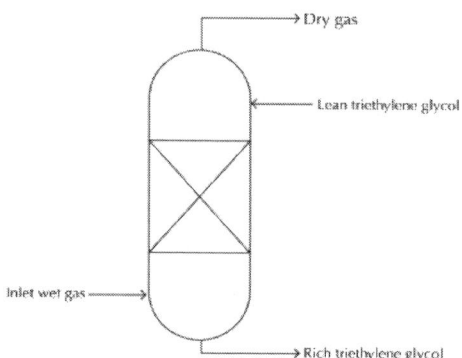

Figure 1: Schematic of the dehydrator.

The entering wet gas enters the bottom of the absorber and flows up counter currently against the lean triethylene glycol, which enters at the top of the absorber. The triethylene glycol absorbs water vapour from the wet gas as it flows down the column and leaves the bottom of the column rich in water, whereas dry gas leaves from the top of the dehydrator. Therefore, the mass diffusion principles governing this operation will be used in developing the mathematical models for the dehydrator. The models would be developed using the principle of conservation of mass and energy to predict the variation of water content in the gas and the variation of temperature of the gas and triethylene glycol across the height of the dehydrator.

MODEL FORMULATION/ ASSUMPTIONS

The following assumptions were made to develop the model.
- Since the column requirement is a diameter ≤ 0.65 m, a packing height of ≤6 m and the fluid are corrosive coupled with a

minimum pressure drop across the column, and packed column is preferred to plate column [14, 15].
- The absorber is well lagged; hence, the heat losses are negligible.
- Since the water vapour in the wet gas is the only diffusing component, no diffusing term would be considered for the liquid phase.
- The effect of change in total molar flow rate is ignored, and an average value is assumed constant [16, 17].
- Vapour-liquid equilibrium relationship is described using Raoult's law and Antoine's equation used for calculation of vapour pressure [18].

Model Development

Material balance (gas phase): Figure 2 shows the elemental packed volume and its flow.

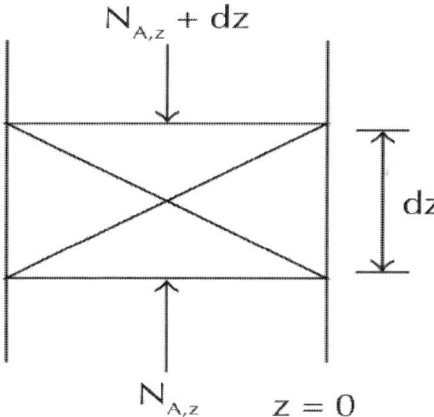

Figure 2: Elemental packed volume.

Consider a homogeneous medium consisting of wet gas (A) and nondiffusive triethylene glycol (B). Let the packed bed be stationary (i.e., the molar average velocity of the mixture is zero), and the mass transfer may occur only by diffusion.

Now consider a differential control volume $d_x\, d_y\, d_z$.

Mass Balance

A general equation can be derived for a binary mixture of wet gas and nondiffusive triethylene glycol for diffusion and convection that also includes terms for unsteady-state diffusion and chemical reaction. Making the material balance on the wet gas on an element of d_x, d_y and d_z fixed in space and shown in Figure 2,

$$-\left[\frac{\partial}{\partial x}(N_{A,x}) + \frac{\partial}{\partial y}(N_{A,y}) + \frac{\partial}{\partial z}(N_{A,z})\right] = \frac{\partial C_A}{\partial t}. \quad (1)$$

For a packed column that is, stationary media, applying Fick's law (1) reduces to

$$-\left[\frac{\partial}{\partial x}\left(D_{AB}\frac{\partial C_A}{\partial x}\right) + \frac{\partial}{\partial y}\left(D_{AB}\frac{\partial C_A}{\partial y}\right) + \frac{\partial}{\partial z}\left(D_{AB}\frac{\partial C_A}{\partial z}\right)\right] = \frac{\partial C_A}{\partial t}. \quad (2)$$

If D_{AB} is constant, (2) becomes

$$-\left[\frac{\partial^2 C_A}{\partial x^2} + \frac{\partial^2 C_A}{\partial y^2} + \frac{\partial^2 C_A}{\partial z^2}\right] = \frac{1}{D_{AB}}\frac{\partial C_A}{\partial t}. \quad (3)$$

Since the absorber is in vertical position,

$$-\frac{d^2 C_A}{\partial x^2} = \frac{\partial^2 C_A}{\partial y^2} = 0. \quad (4)$$

Equation (3) now becomes

$$\frac{\partial^2 C_A}{\partial z^2} = \frac{1}{D_{AB}}\frac{\partial C_A}{\partial t}. \quad (5)$$

But

$$C_A = C_{AO}(1 - y_A).\tag{6}$$

Differentiating (6),

$$dC_A = -C_{AO}dy_A,\tag{7a}$$

$$d^2C_A = -C_{AO}d^2y_A.\tag{7b}$$

Substituting (7a) and (7b) into (5) gives

$$\frac{\partial y_A}{\partial t} = D_{AB}\frac{\partial^2 y_A}{\partial z^2}.\tag{8}$$

The model equation (8) can be used to predict the variation of water content of gas along the column height at different residence times.

Energy Balance

The energy balance will be carried out using the principle of conservation of energy for both the gas and the liquid triethylene glycol. The glycol enters the column at a higher temperature, transferring some amount of heat to the gas, and hence gas phase energy balance is included.

Energy Balance for the Gas Phase

Figure 3 depicts the hypothetical representation of the differential element for energy balance of the gas phase within the packing height, where T_{og} and T_g are the inlet and outlet temperature of the gas, q_z and

d_{qz} are the inlet quantity of heat and outlet quantity of heat from the packing space (d_z), and d_z is the incremental height of the packing space.

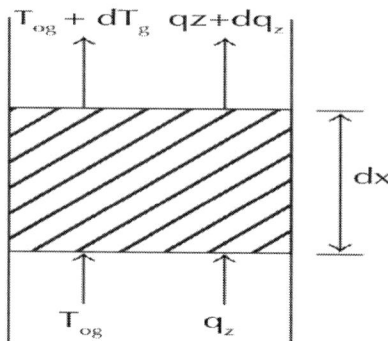

Figure 3: Hypothetical representation of the energy balance within the differential packing height.

Taking cognizance of the conduction of heat axially up the column due to molecular diffusion, the energy balance of the differential element applying the conservation principle gives

$$\frac{\partial T_g}{\partial t} = \frac{\dot{m}}{AC_{Ag}} \frac{\partial T_g}{\partial z} + \frac{K_g \partial^2 T_g}{C_{Ag} C_{pg} \partial z^2} + \frac{Q}{A dz C_{Ag} C_{pg}}, \tag{9}$$

where C_{pg} is the specific capacity of water vapour in the gas stream, q_z is the heat flux in the -direction due to molecular conduction by Fourier's law, A is the area of the packing space, and Q is the amount of heat transferred from the lost glycol to the gas steam. The heat transfer at constant pressure is given by Vuthaluru and Bahadori [19] as

$$Q = LC_{PL} dT_L, \tag{10}$$

Where L is the molar flow rate of triethylene glycol in mol/s and C_{PL} and dT_L are the heat capacity and temperature difference of the liquid glycol.

Recall from dimensional analysis that

$$\frac{k_g}{C_{Ag}C_{Pg}} = \alpha_g, \qquad (11)$$

where α_g is the thermal diffusivity of water vapour in m²/S. Substituting (10) and (11) into (9) results into

$$\frac{\partial T_g}{\partial t} = -\frac{\dot{m}}{AC_{Ag}}\frac{\partial T_g}{\partial z} + \alpha\frac{\partial^2 T_g}{\partial z^2} + \frac{LC_{PL}}{AC_{Ag}C_{Pg}}\frac{\partial T_L}{\partial z}. \qquad (12)$$

Let $\gamma = -\dot{m}/AC_{Ag}$ and $\beta = (LC_{PL}/AC_{Ag}C_{Pg})(\partial T_L/\partial z)$, giving

$$\frac{\partial T_g}{\partial t} = \gamma\frac{\partial T_g}{\partial z} + \alpha\frac{\partial^2 T_g}{\partial z^2} + \beta. \qquad (13)$$

Energy Balance for the Liquid Phase

Figure 4 depicts the hypothetical representation of the inlet and outlet flow into and out from the differential packing bed in the column.

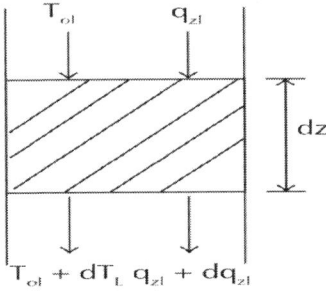

Figure 4: Schematic energy balance for liquid phase.

Similarly the energy balance for the liquid phase is made using the principles of conservation of energy taking cognizance that the triethylene glycol flows from the top to the bottom of the column to obtain

$$\frac{\partial T_L}{\partial t} = \frac{-\dot{m}_L}{AC_{AL}}\frac{\partial T_L}{\partial z} + \alpha_L \frac{\partial T_L^2}{\partial z^2}, \tag{14}$$

Where $\alpha_L = k_L / C_{AL} C_{PL}$ is the thermal diffusivity in the liquid phase (triethylene glycol) in m²/s.

Equations (8), (13), and (14) constitute the mass balance for the water content in feed gas and the energy balances for gas temperature and TEG temperature variations, respectively, in the absorber.

Operational Parameter and Solution Techniques

Operational Parameters

The input and output operating conditions and the physical properties of the wet gas and glycol (density, molecular weight, and molar volume mass and thermal diffusivity) were estimated from an industrial plant [3,20] and are presented in Tables 1, 2, and 3.

Table 1: Inlet conditions [20]

Components	Input streams			
	Gas stream		Glycol stream	
	Weight %	Mol %	Weight %	Mol %
TeG	—	—	99.51	96.054
H2O	0.17	0.187	0.49	3.946
Gas	99.83	99.813	—	—
Total	100.00	100.00	100.00	100.00

| Temperature °C | 50 | 55 |

Table 2: Outlet conditions [20]

Components	Output streams			
	Gas stream		Glycol stream	
	Weight %	Mol %	Weight %	Mol %
TeG	—	—	95.36	71.127
H2O	0.01	0.011	4.04	28.873
Gas	99.99	99.989	—	—
Total	100.00	100.00	100.00	100.00
Temperature °C	51.3		51	

Table 3: Physical properties of components [3]

Properties	TEG	H2O	GAS
Molar mass	150.17	18.02	19.83
Molar volume, m3/Kmol		0.01813	—
Mass diffusivity, m2/S		3.80×10-10	
Thermal diffusivity, m2/S		2.338×10-5	
Density, Kg/m3	1125	1000	

Empirical Evaluation of Mass Diffusivity. The mass diffusivity of water vapour in triethylene glycol (TEG) is evaluated using the formula [3, 21]

$$D_{12} = 1.1728 \times 10^{-16} T \frac{(x^2 M^2)^{1/2}}{\mu_2 V_1^{0.6}}, \qquad (15)$$

where subscript 1 represents the water vapour in the gas, and subscript 2 represents triethylene glycol, where T=50°C = 323.15 K and at T=20°C, μ_2=0.01355515 Pasec.

Solvent Association Parameters. $X_2 = 1$ for (TEG), $V_1=0.0183$ m³/Kmol, $M_2=150.17$ and Kg/Kmol.

Substitution of these values into (15) gives

$$D_{12} = 3.80 \times 10^{-10} \text{ m}^2/\text{s}. \tag{16}$$

Inlet Gas and Glycol Water Content. The inlet gas and glycol water content (in weight %) were obtained from plant operating data and were analytically converted to mol% (assuming binary mixture) using the relations. Consider

$$y_1 = \frac{x_1/m_1}{x_1/m_1 + x_2/m_2},$$

$$y_2 = 1 - y_1, \tag{17}$$

where x_1 and x_2 are concentrations of gas and glycol in wt.%, respectively, y_1 and y_2 are their respective mol%, and m_1 and m_2 are their molecular weight.

Solution Techniques

A numerical solution based on the finite divided difference scheme was developed and keyed into MATLAB program to solve the condensed models for gas water content, gas temperature, and TEG temperature variations given in (8), (13), and (14), respectively.

The developed finite divided difference schemes yield finite grids and computational stencils representing the Y_A, T_g, and T_L, from which Boundary conditions were specified according to "Dirichlet BC." These boundary conditions and initial conditions are given below.

For the gas water content model,

$$y_A(z_o, t) = y_{Ao}, \quad \text{that is, for } z = z_o = 0, \ 0 \leq t \leq t_m,$$

$$y_A(z_n, t) = y_{Af}, \quad \text{that is, for } z = z_n = H, \ 0 \leq t \leq t_m, \tag{18}$$

where Y_{Ao} and Y_{Af} are initial and final water content in gas stream, respectively.

The above boundary conditions explain that the initial gas water content is fixed at the inlet point of absorber column (Z=0) and change with varying values of residence time ranging from 0 to t_m. More so, the final gas water content is established at the outlet point of the absorber column (Z=H) for changing residence time values ranging from 0 to t_m.

The initial condition is

$$y_A(z, t_o) = 0, \quad \text{that is, for } t = t_o = 0, \ 0 \leq z \leq H. \tag{19}$$

implies that the gas water content is established at only zero residence time for varying absorber column height ranging from bottom to top of column.

For the gas temperature model,

$$T_g(z_o, t) = T_{gi}, \quad \text{that is, for } z = z_o = 0, \ 0 \leq t \leq t_m,$$

$$T_g(z_n, t) = T_{go}, \quad \text{that is, for } z = z_n = H, \ 0 \leq t \leq t_m, \tag{20}$$

where T_{gi} and T_{go} are inlet and outlet gas temperatures, respectively.

The above boundary conditions explain that the initial gas temperature is fixed at the inlet point of absorber column (Z=0) and changes with varying values of residence time ranging from 0 to t_m. More so, the final gas temperature is established at the outlet point of the absorber column (Z=H) for changing residence time values ranging from 0 to t_m.

The initial condition is

$$T_g(z, t_o) = 0, \quad \text{that is, for } t = t_o = 0, \ 0 \leq z \leq H. \tag{21}$$

This implies that the gas temperature is established at only zero residence time for varying absorber column height ranging from bottom to top of column.

For the TEG temperature model,

$$T_L(z_o, t) = T_{Li}, \quad \text{that is, for } z = z_o = 0, \; 0 \leq t \leq t_m,$$
$$T_L(z_n, t) = T_{Lo}, \quad \text{that is, for } z = z_n = H, \; 0 \leq t \leq t_m, \quad (22)$$

where T_{Li} and T_{Lo} are inlet and outlet TEG temperatures, respectively.

The above boundary conditions explain that the initial TEG temperature is fixed at the inlet point of absorber column (Z=0) and changes with varying values of residence time ranging from 0 to t_m. More so, the final TEG temperature is established at the outlet point of the absorber column (Z=H) for changing residence time values ranging from 0 to t_m.

The initial condition is

$$T_L(z, t_o) = 0, \quad \text{that is, for } t = t_o = 0, \; 0 \leq z \leq H. \quad (23)$$

This implies that the TEG temperature is established at only zero residence time for varying absorber column height ranging from bottom to the top of column.

RESULTS AND DISCUSSION

Table 4 shows the comparison between plant data and predictions from model (see (8), (13), and (14)), indicating that the predicted results agree reasonably well with the plant data. These results show a deviation ranging from 3.41 to 9.18 percent.

Table 4: Comparison between plant data and model predictions

Process parameter	Model prediction	Plant data	% deviation
Final gas water content	7.93×10-7	7.24×10-7	8.65

| Gas outlet temperature (°C) | 44.52 | 43 | 3.41 |
| TEG outlet temperature (°C) | 48.45 | 44 | 9.18 |

Profiles presented and discussed here will subsequently reveal the following: variations of gas water content with time and axial height of packing in the column, variation of temperature of triethylene glycol (TEG) with column height at different thermal diffusivities, variation of temperature of gas with column height at different residence times, variation of gas water content across column height at different mass diffusivities, and variation of temperature of triethylene glycol across the column height at different residence times.

Variation of Water Content of Gas with Column Height at Different Residence Times

It can be deduced from Figure 5 that the water content of the gas reduces as the gas moves from the bottom of the column to the top. It can also be deduced that the higher the residence time of the gas in the column, the higher the rate of removal of the water vapour from the gas. This holds true since a relatively smaller time is needed to establish equilibrium between the water vapour in the gas and that in the liquid phase [10, 22]. This means that as the residence time increases, say, to 200 seconds, the water vapour returns to the vapour phase again implying that the water content in gas increases. It can also be deduced from Figure 5 that, at a height of approximately 7 m and above, the gas water content variation becomes steady.

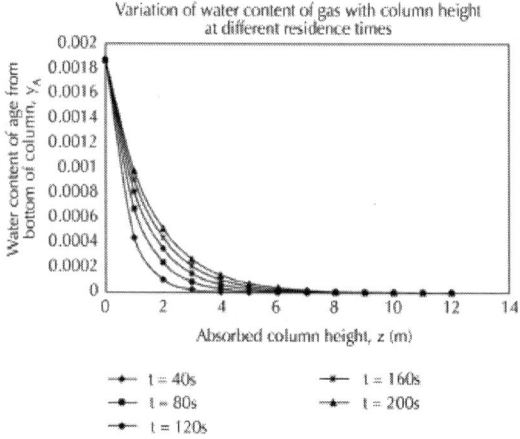

Figure 5: Variation of gas water content (mole fraction) from bottom of column.

In addition, the solutions of the model will be represented as a three-dimensional surface plot in Figure 6. The purpose of the surface plots is to visualize the propagation of the gas water content in time and space and to make conclusions based on the subsequent trends. The surface plots are not intended to give the exact numerical values but for visualization.

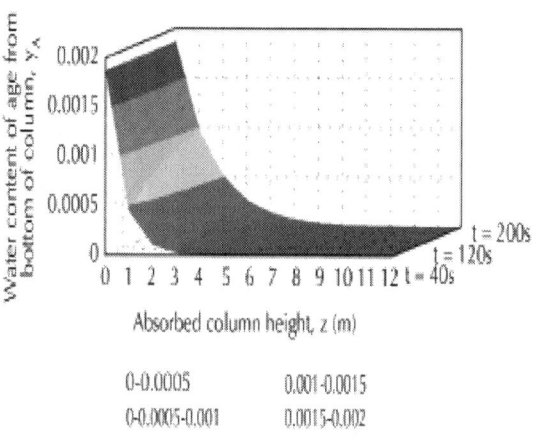

Figure 6: Surface plot showing gas water content propagation along the column.

The natural gas propagates from the base of the absorber and initially holds a water concentration of 0.187 mol%. The low resistance in the gas bulk will cause the gas and liquid bulk phases to reach chemical equilibrium virtually instantaneous. The steep transient observed at the lower part of the column confirms the trend illustrated by Figure 6; it is also in agreement with plant data. Also, as operation proceeds half way up the column, the absorption of water from natural gas becomes numerically insignificant and remains practically constant.

Variation of Temperature of Triethylene Glycol with the Column Height at Different Residence Times

In Figure 7, the temperature of the absorbing solvent, triethylene glycol (TEG), reduces gradually as it travels from the top of the column to the bottom. Initially, triethylene glycol enters the column at a temperature of 50°C and leaves at 46.5–48°C depending on the residence time. It is observed that the temperature change becomes smaller as the residence time increases, resulting into a very steep slope at time = 200 seconds. This is obviously because more water vapour has been absorbed at higher residence time.

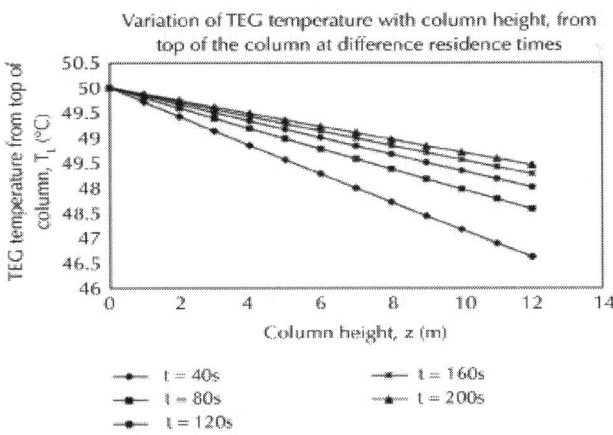

Figure 7: Variation of temperature of triethylene glycol down the column.

The surface plot in Figure 8 further visualizes the TEG temperature reduction at varying residence times. The largest TEG temperature change is achieved at a residence time of 40 to 80 seconds, while a little change is achieved at 160 to 200 seconds.

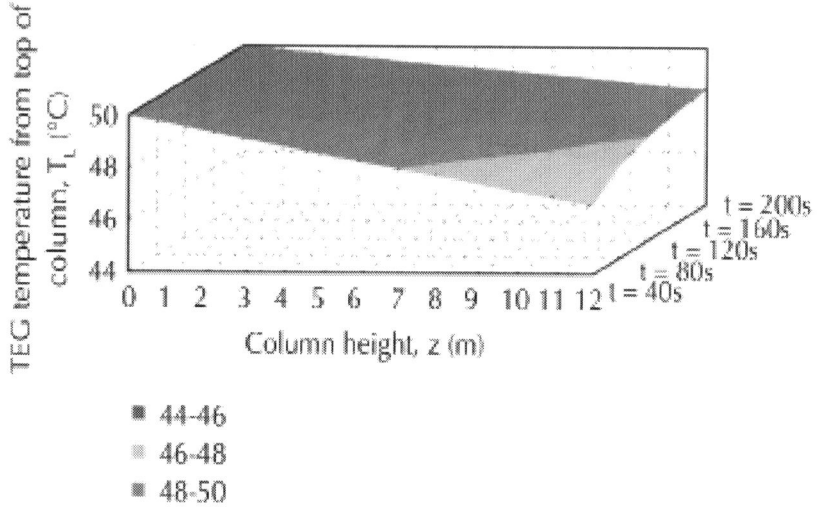

- 44-46
- 46-48
- 48-50

Figure 8: Surface plot showing triethylene glycol temperature propagation along the column.

Variation of Temperature of Gas with Column Height at Different Residence Times

The gas enters the column at a temperature of 42.5°C and increases very slowly until it leaves the column at a slightly higher temperature of 46.5°C. However, at a residence time t=40 seconds and at column height z=12 m, the outlet temperature of gas is approximately 44.5°C. It can be noticed in Figure 9 that the temperature becomes lower at higher residence time. This implies that if the residence time is reduced further, the required outlet gas temperature would be achieved at approximately 120 seconds. It should be noted that the increase in residence time results from the transfer of heat from the liquid stream to the gas stream.

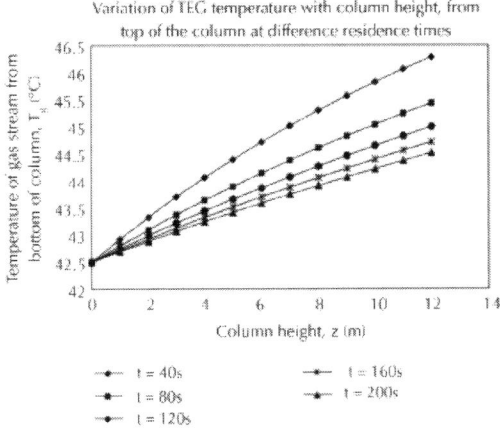

Figure 9: Variation of the temperature of the gas stream with column height from column bottom.

The surface plot in Figure 10 reveals that the gas temperature variation is not widely distributed. It can be further deduced that the lowest trend in gas temperature is attained at residence time of 120 seconds, while the largest change is obtained at residence time of approximately 40 to 80 seconds.

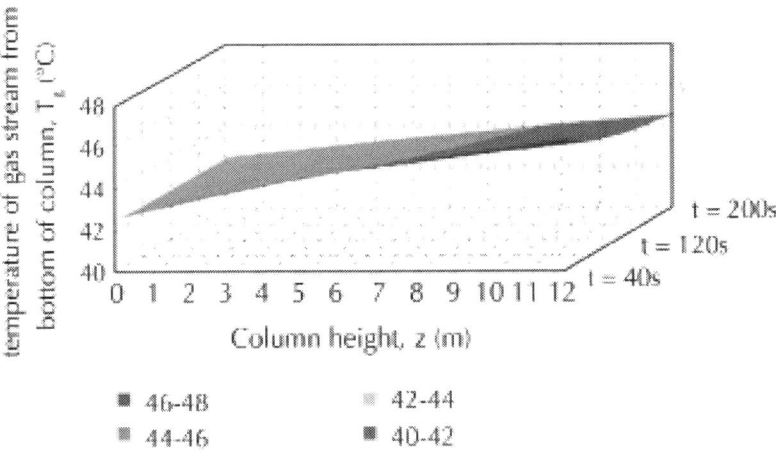

Figure 10: Surface plot showing gas temperature variation along the column.

Variation of Water Content of Gas with Column Height at Different Mass Diffusivities

The mass diffusivity is the property of a material that determines the rate at which a given component is transferred across a concentration gradient. This property is a vital parameter in this work. From Figure 11, it is evident that, given a fixed time of 40 seconds, at higher mass diffusivities, the rate of transfer of water vapour from gas to the liquid stream decreases slightly as up the column. Also, as the mass diffusivity reduces, the rate of transfer of water vapour from gas to the triethylene glycol stream increases sharply. This implies that the mass diffusion coefficient of the gas through the TEG should be as low as $\leq 3.80 \times 10^{-10}$ m²/s for optimal absorption.

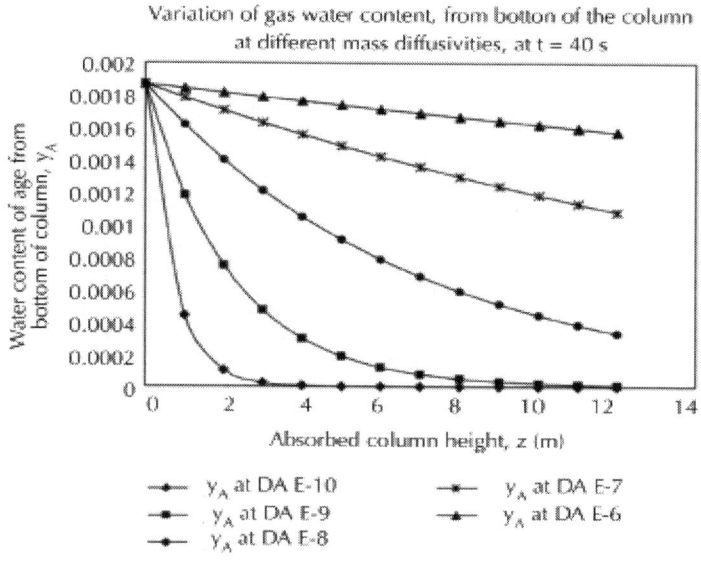

Figure 11: Variation of water content of gas at different mass diffusivities at seconds.

The surface plot in Figure 12 shows that the gas water content variation widely spreads across the column height at different mass diffusivities. Similar final gas water content values are obtained at mass diffusivities of 3.80×10^{-9} m²/s and 3.80×10^{-10} m²/s.

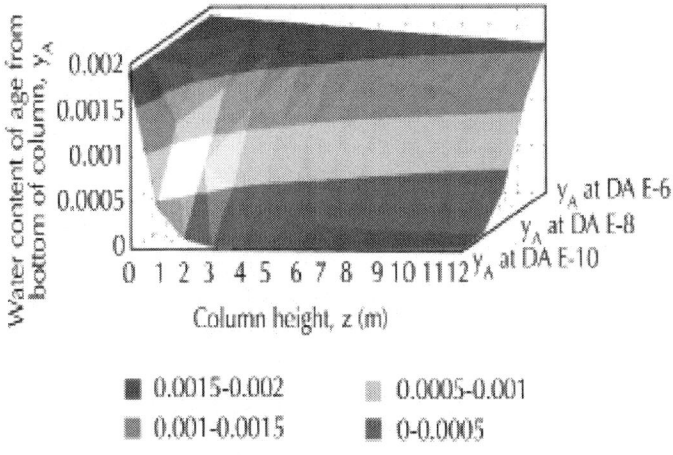

Figure 12: Surface plot showing gas water content variation at different mass diffusivities when time = 40 seconds.

Variation of Temperature of Triethylene Glycol (TEG) with Column Height at Different Thermal Diffusivities

Thermal diffusivity is the property of a material which describes the rate at which heat flows through the material. Water vapour, being a better heat transfer agent, has a value of 2.336×10^{-5} m^2/s, whereas liquid water has a value of 1.4×10^{-5} m^2/s. From Figure 13, it is observed generally that the temperature of the solvent triethylene glycol decreases sharply down the column as the thermal diffusivity decreases. This decrease is faster as the thermal diffusivity decreases, resulting in a steep slope at the lowest thermal diffusivity of 2.338×10^{-5} m^2/s. It is imperative to note that these plots were taken at a residence time of t=40 seconds.

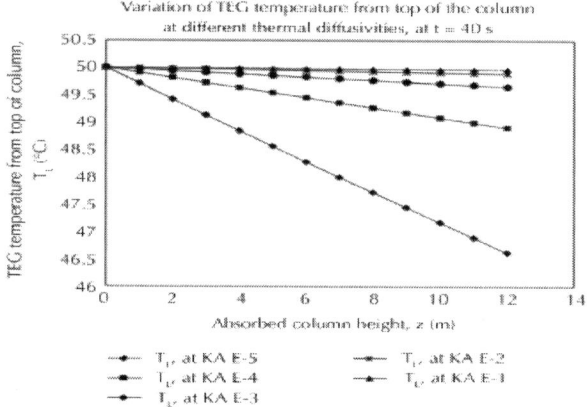

Figure 13: Variation of temperature of TEG with thermal diffusivities at t=40 seconds.

In addition, the surface plot in Figure 14 clearly visualizes the TEG temperature decrease along the column and at varying thermal diffusivities. A relatively large TEG temperature change exists between thermal diffusivities of 2.338×10^{-4} and 2.338×10^{-5} m²/s; conversely, there is a insignificant temperature change at 2.338×10^{-1} m²/s.

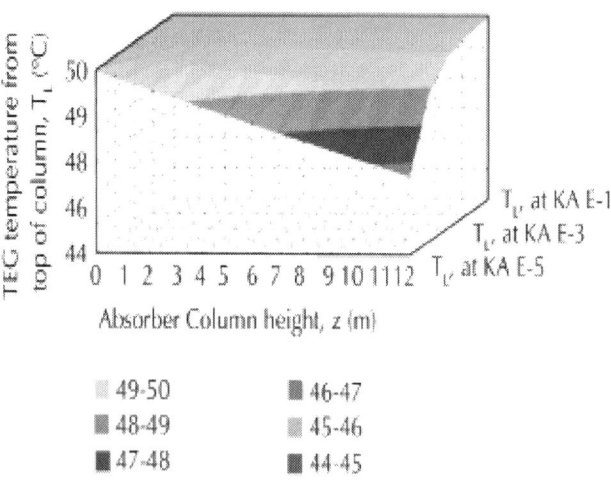

Figure 14: Surface plot of TEG temperature at different thermal diffusivities at seconds.

CONCLUSIONS

Mathematical models of the absorber of a glycol dehydration facility were developed using the principles of conservation of mass and energy. The models could predict the variation of the water content of gas in mole fraction and the gas and liquid (TEG) temperatures across the packing height. The models developed contain contributions from bulk and diffusion flows. The models were validated using the initial conditions from a functional industrial TEG unit in Nigeria to ascertain if the outlet conditions predicted by the models meet the industrial plant outlet values.

REFERENCES

1. J. M. Campbell, R. N. Maddox, L. F. Sheerar, and J. H. Erbar, Gas Conditioning and Processing, vol. 3 ofCampbell Petroleum Series, Campbell Petroleum, Norman, Okla, USA, 1982.
2. J. M. Campbell, Gas Conditioning and Processing, vol. 2, Campbell Petroleum Series, Norman, Okla, USA, 7th edition, 1992.
3. R. H. Perry and D. W. Green, Perry's Chemical Engineers' Handbook, McGraw Hil, New York, NY, USA, 7th edition, 1999.
4. N. A. Darwish and N. Hilal, "Sensitivity analysis and faults diagnosis using artificial neural networks in natural gas TEG-dehydration plants," Chemical Engineering Journal, vol. 137, no. 2, pp. 189–197, 2008.
5. A. Bahadori, H. B. Vuthaluru, and S. Mokhatab, "Analyzing solubility of acid gas and light alkanes in triethylene glycol," Journal of Natural Gas Chemistry, vol. 17, no. 1, pp. 51–58, 2008.
6. A. Bahadori and H. B. Vuthaluru, "Simple methodology for sizing of absorbers for TEG (triethylene glycol) gas dehydration systems," Energy, vol. 34, no. 11, pp. 1910–1916, 2009.
7. C. U. Ikoku, Natural Gas Production Engineering, Kreiger Publishing, Malabar, Fla, USA, 1992.
8. Gas Processors Suppliers Association (GPSA), Engineering Data Book, chapter 20, Gas Processors Suppliers Association, Tusla, Okla, USA, 11th edition, 1998.

9. B. M. Jaćimović, S. B. Genić, D. R. Djordjević, N. J. Budimir, and M. S. Jarić, "Estimation of the number of trays for natural gas triethylene glycol dehydration column," Chemical Engineering Research and Design, vol. 89, no. 6, pp. 561–572, 2011.
10. J. F. Richardson, J. H. Harker, and J. R. Backhurst, Coulson Richardsons Chemical Engineering: Chemical Engineering Design, vol. 6, Elsevier, New Delhi, India, 3rd edition, 2002.
11. S. Chapra and R. P. Canale, Numerical Methods for Engineers, McGraw Hill International Edition, 6th edition, 2009.
12. A. Ahmed, T. Paitoon, and I. Raphael, Chemindix, ccu/09, International Test Centre for Carbon Dioxide Capture (ITC), Faculty of Engineering, University of Regina, Regina, Canada, 2007.
13. S. M. Peyghambarzadeh and M. Jafarpour, "Impact of thermodynamic model on the simulation of natural gas dehydration unit," in Proceedings of the 6th National-Student Chemical Engineering Congress, pp. 1–10, University of Isfaham, 2006.
14. S. Max, D. T. Klaus, and E. W. Ronald, Plant Design and Economics for Chemical Engineers, McGraw Hill International, New York, NY, USA, 5th edition.
15. L. Mei and Y. J. Dai, "A technical review on use of liquid-desiccant dehumidification for air-conditioning application," Renewable and Sustainable Energy Reviews, vol. 12, no. 3, pp. 662–689, 2008
16. W. L. McCabe, J. Smith, and P. Harriott, Unit Operations of Chemical Engineering, McGraw-Hill, New York, NY, USA, 7th edition, 2005.
17. D. M. Himmelblau, Basic Principles and Calculate in Chemical Engineering, Prentice-Hall, New Delhi, India, 6th edition, 2005.
18. Y. S. Choe, Regrowns dynamic models of distillation coloums [M.S. thesis], Lehigh University, Bethlehem, Pa, USA, 1985.
19. H. B. Vuthaluru and A. Bahadori, "A new method for prediction of absorption/stripping factor," Computer and Chemical Engineering, vol. 34, pp. 1731–1736, 2010.
20. Shell Petroleum Development Company (SPDC) of Nigeria, "Gbaran Ubie Integrated Oil & Gas Development Project,

Gas Dehydration & Glycol Regeneration Packages," Operating Manual, 2007.

21. P. Gandhidasan and A. A. Al-Mubarak, "Dehydration of natural gas using solid desiccants," Energy, vol. 26, no. 9, pp. 855–868, 2001.

22. J. F. Richardson, J. H. Harker, and J. R. Backhurst, Coulson & Richardson›s Chemical Engineering: Particle Technology and Separation Process, vol. 2, Elsevier, New Delhi, India, 5th edition, 2002.

Chapter 5

A Field Study on Simulation of CO_2 Injection and ECBM Production and Prediction of CO_2 Storage Capacity in Unmineable Coal Seam

Qin He, Shahab D. Mohaghegh, and Vida Gholami

Department of Petroleum and Natural Gas Engineering, West Virginia University, Morgantown, WV 26505, USA

ABSTRACT

CO_2 sequestration into a coal seam project was studied and a numerical model was developed in this paper to simulate the primary and secondary coal bed methane production (CBM/ECBM) and carbon dioxide (CO_2) injection. The key geological and reservoir parameters, which are germane to driving enhanced coal bed methane (ECBM) and CO_2 sequestration processes, including cleat permeability, cleat porosity, CH_4 adsorption time, CO_2 adsorption time, CH_4 Langmuir isotherm, CO_2 Langmuir isotherm, and Palmer and Mansoori parameters, have been analyzed within a reasonable range. The model simulation results showed good matches for both CBM/ECBM production and CO_2 injection compared with the field data. The history-matched model was used to estimate the total CO_2 sequestration capacity in the field. The model forecast showed that the total CO_2 injection capacity in the coal seam could be 22,817 tons, which is in agreement with the initial estimations based on the Langmuir isotherm experiment. Total CO_2 injected in the first three years was 2,600 tons, which according to the model has increased methane recovery (due to ECBM) by 6,700 scf/d.

INTRODUCTION

Fossil fuels are currently playing a significant role in the whole world's energy supply. However, its damage to the environment, especially the CO_2 emission resulting in the greenhouse effect, has gotten more and more attention. At present, several geological CO_2 sequestration technologies, such as CO_2 injection into saline aquifer, CO_2-EOR, CO_2-ECBM, and so forth, have been studied to minimize the CO_2 release into the atmosphere, and these projects have been operating all over the world [1–6]. Studies have shown that unmineable coal seams (seams too deep or too thin to be mined economically) are pretty attractive as one of the promising options for CO_2 sequestration because of their large CO_2 sequestration capacity, long time CO_2 trapping, and extra enhanced coal-bed methane (ECBM) production benefits [1, 7–10]. Field experience with CO_2 injection into coal seam is limited, although field tests are planned or are being conducted in the USA, Canada, Poland, Australia, and Japan [3].

However, unlike conventional reservoirs, gas flow in the coal seams can cause the cleat permeability and porosity variation during the injection/production process. Once gas is injected and adsorbed on the coal matrix, the matrix will swell, and correspondently decrease the cleat permeability and porosity [11, 12]. Due to its special features and the nature of gas retention in CBM reservoirs, simulating the production and injection will have more complexity compared to conventional resources.

Similar to conventional naturally fractured reservoirs, coal is characterized as a dual-porosity system consisting of matrix and cleat, in which majority of the gas is stored within the coal matrix by a process of adsorption and a small amount of free gas exists in the cleats or fractures [13]. Once CO_2 is injected into the coal seam, it will be held by coal surface because of its higher affinity to the coal matrix than methane, and then displaces the methane to boost extra natural gas production. Figure 1 shows a schematic representation of the CO_2 sequestration-ECBM process. It is estimated by laboratory measurements that this process, known as CO_2-enhanced coal bed methane, can store twice as much CO_2 as the methane desorbed or even more [14].

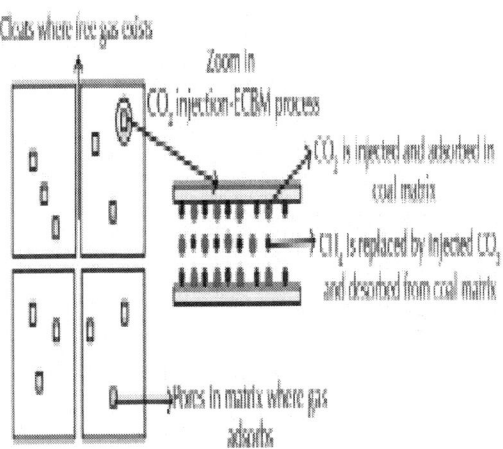

Figure 1: A schematic representation of CO_2 sequestration-ECBM production.

The entire gas flow mechanism can be summarized in three steps: (1) desorption: once free gas or water is produced from fracture systems in coal seams, pressure starts to be released, then the adsorbed gas will be desorbed from the matrix surface, which can be described by Langmuir isotherm equation; (2) diffusion: due to the gas molecular concentration difference, gas will diffuse from matrix surface to cleats/micro-pores; Darcy's flow: gas in the cleats and natural fractures will flow to the wellbore by Darcy's flow [15]. Recently, the numerical reservoir simulator have become the most popular tool to predict coal seam performance and provides a good understanding of gas flow from the reservoir to the wellbore [16].

Langmuir Isotherm

The gas adsorption/desorption process can be described by the typical formulation of Langmuir isotherm:

$$V(P) = \frac{V_L P}{P_L + P}. \tag{1}$$

As shown in Figure 2, Langmuir volume (V_L) is the maximum amount of gas that can be adsorbed on a piece of coal at infinite pressure. Langmuir pressure (P_L) is the pressure at which the Langmuir volume can be adsorbed. V (P) is the amount of gas at different pressure, also known as gas content (scf/ton). Whenever the Langmuir volume and Langmuir pressure are known, the adsorbed gas amount can be calculated at any pressure.

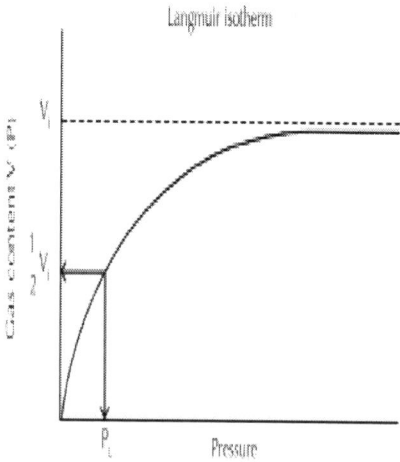

Figure 2: Langmuir isotherm function.

Diffusion

Diffusion is the fact that particles move/spread from high concentration to low concentration region. Diffusion of gas out of the coal matrix can be expressed by a simple diffusion equation. The diffusion process in coal seams can be described by either diffusion coefficient or coal desorption time input in the simulator [16]:

$$\frac{\partial C}{\partial t} = \frac{1}{\tau}\left[\overline{C} - C(P_f)\right].$$

(2)

Coal Shrinkage and Swelling

One of the unique characteristics of coal seam is the phenomenon of pressure dependent permeability. As the production from the reservoir takes places, two distinct phenomena occur. First, the reservoir pressure declines, which causes the pressure in the fractures to decline as well, which in turn leads to an increase in the effective stress within the cleats causing the cleats to be more compactable, so the cleat permeability will decrease. At the same time, the gas that has been desorbed is coming out of the matrix, which causes the matrix to shrink and the cleats to open-up; thereby the cleat permeability will be increased. As a function of the pressure drop, compressibility dominates in early time and shrinkage dominates in the late time [16]. Palmer and Mansoori model [17] is used to simulate the permeability change process during production/injection in this model:

$$\frac{\varnothing}{\varnothing_0} = 1 + C_f\left(\frac{P-P_0}{\varnothing_0}\right) + \frac{\varepsilon_\infty}{\varnothing_0}\left(\frac{K}{M}-1\right)\left(\frac{P}{P+P_L} - \frac{P_0}{P_0+P_L}\right),$$

$$\frac{K}{K_0} = \left(\frac{\varnothing}{\varnothing_0}\right)^3.$$

(3)

PROJECT DESCRIPTION

From 2009, the CO_2 sequestration with ECBM production project began in Marshall County, West Virginia. The objective of this project was to help mitigate climate change by providing an effective and economic way to permanently store CO_2 in un-minable coal seams. In advance of CO_2 injection, four horizontal coalbed methane wells (MH5, MH11, MH18, and MH20) were drilled into the un-minable Upper Freeport coal seam, which are 1,200 to 1,800 feet below the ground. These wells have been producing coalbed methane since 2004. The center located wells (MH18 and MH20) have been converted to CO_2 injection wells since September 2009 [18]. 20,000 short tons are planned to be injected through well MH18 and MH20 in two years.

Several questions come with this project and need to be investigated: how much CO_2 can be stored in this coal seam? How long does the injection process take? Which parameters affect the injection and production the most? These questions could be answered by an effective coal seam model, which was represented by a dual-porosity system to show the fluid flow through both matrix and cleat under the particular conditions in this site. The following assumptions were considered for the modeling and simulation purpose.

- The initial seam pressure is hydrostatic pressure, which is 0.28 psi/ft after water is produced.
- The flow in the coal seam is single phase including only CH_4 and CO_2.
- The fluid flow in the cleat system is a laminar flow due to the larger pore size and it is governed by Darcy's Law, while the flow in the matrix is a diffusional flow due to smaller pore size and governed by Fick's Law.
- Palmer and Mansoori equation is used to allow the natural permeability and porosity to vary as a function of pressure.

In most cases, the actual in situ seam data is unavailable, which leads to the requirements of some assumptions on certain parameters, such as, in this case, matrix/cleat permeability, matrix/cleat porosity, geo-mechanical properties (Young's modulus, Poisson ratio), and so forth. Table 1 summarizes the initial physical parameters in the model.

Table 1: Initial reservoir parameters used in the model

Input parameters	Value	Unit	Input parameters	Value	Unit
Average reservoir depth	1200	ft	Poisson ratio	0.3	
Average formation thickness	4	ft	Young's Modulus	125,000	psia
Fracture spacing I/J/K	0.02	ft	CO_2 Strain	0.0065	
Perm I-Matrix	0.01	md	CH_4 Strain	0.0045	
Perm J-Matrix	0.01	md	Palmer/Mansoori exponent	3	
Perm K-Matrix	0.001	md	CO_2 Langmuir Pressure	240	psia
Perm I-Fracture	0.2	md	CO_2 Langmuir Volume	890	scf/ton
Perm J-Fracture	0.2	md	CH_4 Langmuir Pressure	402	psia
Perm K-Fracture	0.02	md	CH_4 Langmuir Volume	452	scf/ton
Porosity-Matrix	0.004		CO_2 Sorption time	100	days
Porosity-Fracture	0.001		CH_4 Sorption time	100	days
Rock compressibility-Matrix	1.00E-06	1/psi	Rock compressibility-Fracture	1.00E-06	1/psi

HISTORY-MATCHING RESULTS AND DISCUSSION

As indicated before, the CO_2 sequestration-ECBM production project went through three stages: primary methane (CBM) recovery, CO_2 injection, and secondary methane (ECBM) recovery. MH18 and MH20 were firstly performed as production wells from January 2005 to July 2007 with a following two-year shut in period; thereafter, they were transferred into CO_2 injection wells since September 2009. MH5 and MH11 keep on methane production from the all the way from beginning to present. All well productions and injection were simulated starting from the start day until the date the most updated data have been recorded and reported (August 2012 in this paper).

However, different performance of MH18 and MH20 in different time periods introduced a lot of complexity on the history matching process. A key factor should be respected in the history matching: either for initial methane production or the following CO_2 injection, well properties (MH18 or MH20) must stay the same in the model; thereby what was changed is only the operation type.

The results of sensitivity analysis were very valuable in back and forth model parameter adjustment. Sensitivity analysis is known as the study of how the variation (uncertainty) in the output of a mathematical model can be apportioned, qualitatively or quantitatively affected by the change of different variations in the input of the model [19]. Sensitivity analysis of coal modeling properties is widely studied and is addressed that it will be an important tool in future decision making [19–21]. In this case, related coal parameters, including cleat permeability, porosity, CH_4 desorption time, CO_2 desorption time, CH_4 Langmuir volume, CO_2 Langmuir volume, and Palmer and Mansoori parameters have been tested in the model. The comparison of coal physical property influences can be concluded based on the study result as: Young's modulus and Poisson ratio have little effect, while sorption time, cleat permeability, strain, and Langmuir isotherm are the key parameters that affect CH_4 production and CO_2 injection most.

The actual in-seam data for both methane production and CO_2 injection in Upper Freeport coal seam were reported daily as shown in Figure 3. The average minimum bottom hole pressure in production wells is 20 psia, and the average maximum BHP in injection wells is 900 psia. The daily injection rate is set as constraint. The trend could be observed in the production; the methane production rate has clearly increased in MH5 and MH11 after July 2009 due to the CO_2 injection. A gradual decline trend in injection rate can be noticed in the injection wells, especially in MH18, which can be a consequence of the permeability changes occurring during desorption/adsorption process on coal.

A Field Study on Simulation of CO$_2$ Injection and ECBM ...   123

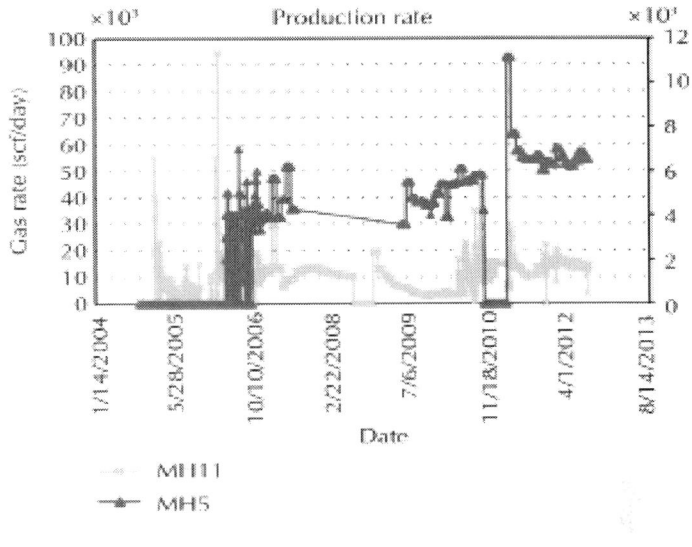

(a)

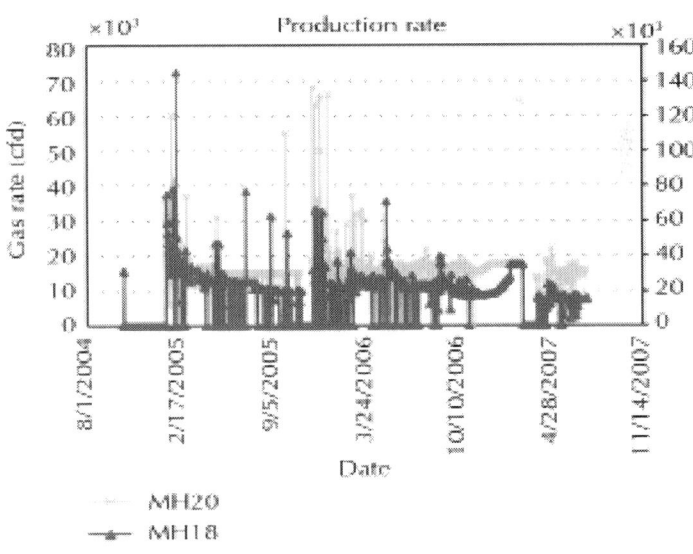

(b)

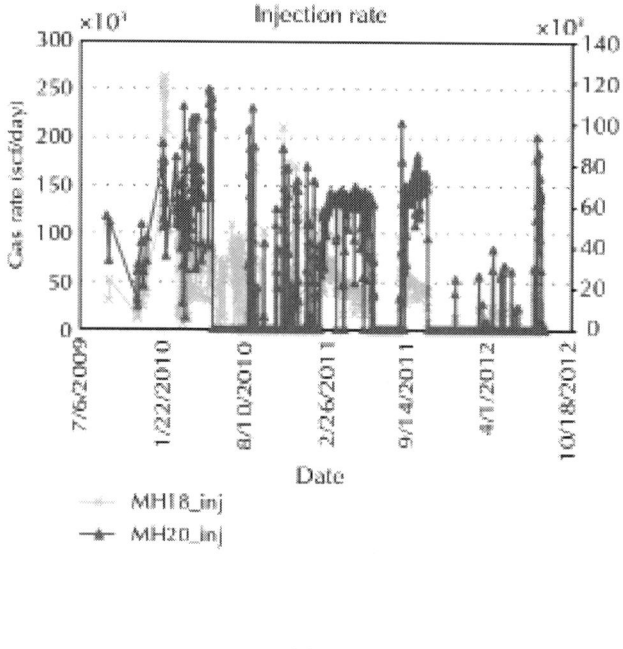

(c)

Figure 3: Actual CH_4 production rate/CO_2 injection rate in Upper Freeport coal seam. (a) CH_4 production rate in MH11 and MH5, (b) CH_4 production rate in MH18 and MH20, (c) CO_2 injection rate in MH18_inj and MH20_inj (MH18 and MH20 after conversion to Injection wells).

No regular tracking pattern of daily rate was observed because of frequent shut-in operations due to weather, equipment damage, or other unpredictable reasons during the injection process. Therefore, cumulative rates are considered to be the history matching target by setting bottom hole pressure as constraints in the model. History matching was performed for six wells, and final existing reservoir properties, including permeability, porosity, Langmuir isotherm parameter, sorption time, and so forth, as appropriate, were determined by history matching. The history matching results are illustrated in Figures 4 and 5 and the coal parameters are listed in Table 2. It is important to note that the degree of component isotherm and sorption time at any given in-situ condition is directly related to the rank of the coal. Values may change in a large range from different coal seams.

Table 2: History matched reservoir parameter setting

Input parameters	Value	Unit	Input parameters	Value	Unit
Average reservoir depth	1200	ft	Poisson ratio	0.3	
Average formation thickness	4	ft	Young's Modulus	125,000	psia
Fracture spacing I/J/K	0.015	ft	CO_2 Strain	0.0025	
Perm I-Matrix	0.01–0.02	md	CH_4 Strain	0.0045	
Perm J-Matrix	0.01–0.02	md	Palmer/Mansoori exponent	3	
Perm K-Matrix	0.001–0.002	md	CO_2 Langmuir Pressure	412	psia
Perm I-Fracture	0.2–0.4	md	CO_2 Langmuir Volume	800	scf/ton
Perm J-Fracture	0.2–0.4	md	CH_4 Langmuir Pressure	628	psia
Perm K-Fracture	0.02–0.04	md	CH_4 Langmuir Volume	652	scf/ton
Porosity-Matrix	0.002–0.004		CO_2 Sorption time	140	days
Porosity-Fracture	0.001–0.002		CH_4 Sorption time	350	days
Rock compressibility-Matrix	1.00E-06	1/psi	Rock compressibility-Fracture	1.00E-06	1/psi

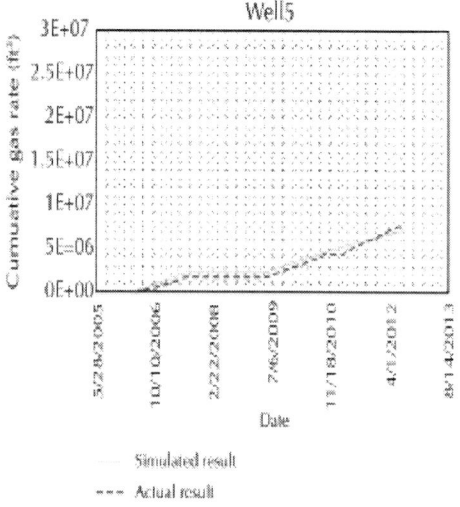

(a)

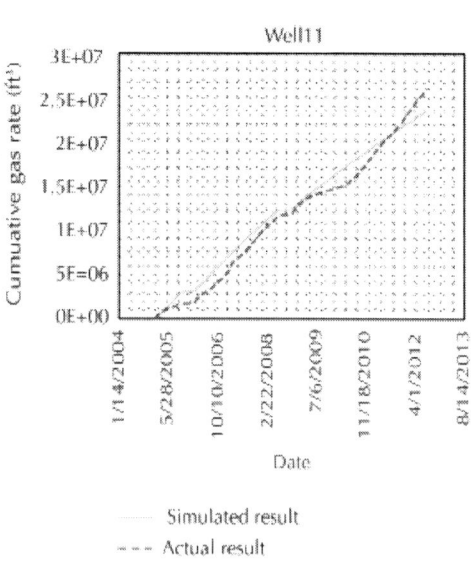

(b)

A Field Study on Simulation of CO_2 Injection and ECBM ...

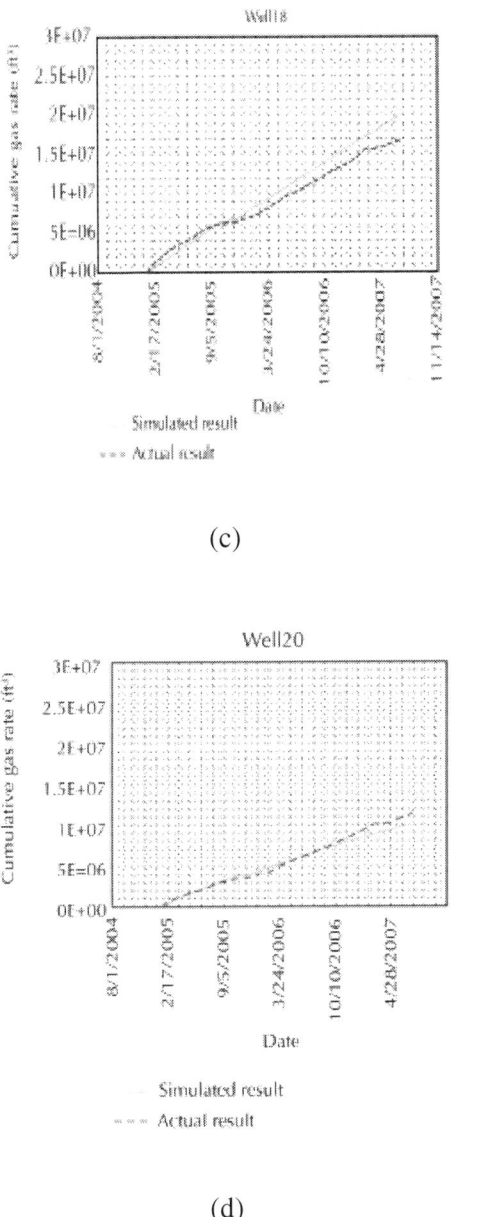

(c)

(d)

Figure 4: CH_4 cumulative production history matching. (a) CH_4 cumulative production in MH5, (b) CH_4 cumulative production in MH11, (c) CH_4 cumulative production in MH18, (d) CH_4 cumulative production in MH20.

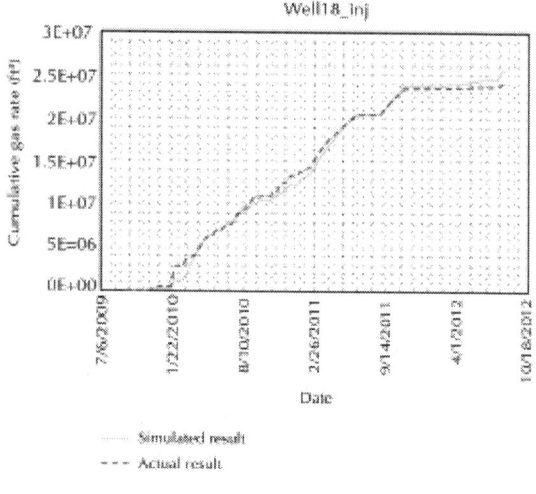

(a)

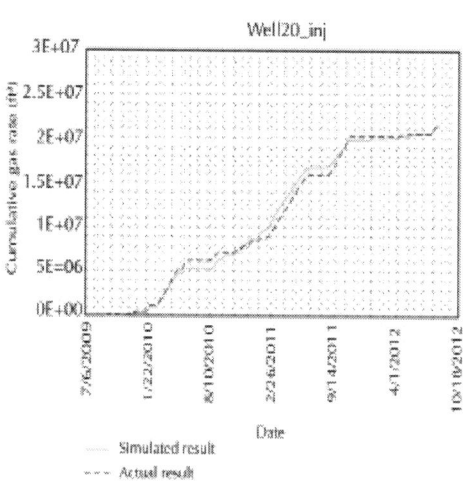

(b)

Figure 5: Cumulative CO_2 injection history matching. (a) Cumulative CO_2 injection in MH18_inj, (b) Cumulative CO_2 injection in MH20_inj.

Figure 4 shows the fairly good history matching result of CH_4 cumulative production for all production wells. Green line and red line represents the simulated result and actual data, respectively. As shown in Figure 4(a), well5 was shut in from July 2007 to April 2009 and October 2010 to March 2011, which can be seen from two short straight lines in red cumulative curves. $7 \times 10^6 ft^3$ CH_4 could be produced from well5 by August 2012 with a stable increase. As illustrated in Figure 4(b), well11 had a short shut-in period of three months; that is why no production increase is shown in October 2005 and from July 2008 to November 2008. Totally, $2 \times 10^7 ft^3$ CH_4 were produced from well11 by August 2012, a sharp build-up could be observed after the start of large CO_2 injection on September 2009, which is because of ECBM production. Figures 4(c) and 4(d) show the cumulative CH_4 production of well18 and well20 from January 2005 to July 2007, respectively, before they were shut-in and transferred to CO_2 injection well. MH18 produced $1.6 \times 10^7 ft^3$ CH_4, while MH20 had a total of $1 \times 10^7 ft^3$ CH_4 production at the end of production period.

Figure 5 shows cumulative CO_2 injection history matching in MH18 and MH20 after they were converted into injection wells. Red dashed line represents actual CO_2 injection data from September 2009 to August 2012, while green line shows simulation results for both wells. Certain plateaus could be seen in the curves during the whole injection periods, which is because of the shut-in times resulting from operational reasons, such as weather affects, equipment damage, and so forth. More CO_2 was injected through well18 (maximum amount of $2.5 \times 10^7 ft^3$ CO_2), compared to $2.5 \times 10^7 ft^3$ CO_2 injection in well20. The total amount of injected CO_2 through MH18 and MH20 has been almost 3,000 tons in the first three years, with an average ECBM increase of an approximation of 6,700 scf/day.

CO_2 SEQUESTRATION CAPACITY IN COAL SEAM

There are four main CO_2 storage mechanisms in coal seams: (a) stratigraphic and structural trapping, (b) hydrodynamic trapping, (c) mineral trapping, and (d) adsorption trapping. In un-mineable coal seams, adsorption trapping is the main sequestration method. This is the process of accumulation of injected gases which is adsorbed on the surface of micropores within the coal matrix. The adsorption capacity will mostly depend upon Langmuir isotherm factors [22]. Figure 6 illustrates the final Langmuir Isotherm in Upper Freeport coal seam in this case.

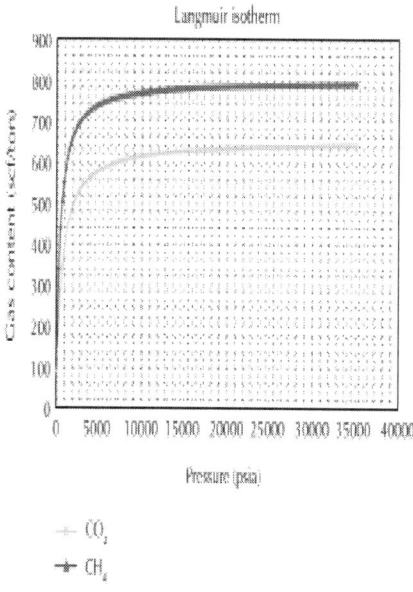

Figure 6: Existing Langmuir isotherm for CO_2 and CH_4 in Upper Freeport Coal seam.

Two assumptions have been made in order to simplify the calculation here.

- No water production data was reported in this case; the coal reservoir was simulated with single phase production with only CH_4 and CO_2.
- Adsorption trapping is the main sequestration method in unmineable coal seam, which was considered as the only storage mechanism without including free gas in the fractures in this study.

The CO_2 adsorption capacity in the coal seam can be calculated as:

$$\text{OGIP} = A \times h \times \rho_b \times G_{ci} = V \times \rho_b \times G_{ci}, \qquad (4)$$

Where V_L = 800 scf/ton, P_L = 412 psia, P = 0.28psi/ft×1200ft = 360 psi, and $V(P) = G_{ci} = V_L P/(P_L + P) = 800 \times 360/(412 + 360) = 373$ scf/ton, where, ρ_b = 85 lbs/ft^3, V = 25,193,558 ft^3, 1 ton = 2,000 lbs, Coal tonnage = 85 ×25,193,558/2,000 = 1,069,466 tons, OGIP = 1,069,466 tons ×373 scf/ton/17,483 ton/scf = 22,817 tons (coal seam volume and coal density were provided and were used directly).

SUMMARY AND CONCLUSIONS

The modeling and history matching process of methane production and ECBM as well as CO_2 injection in a coal bed seam was explained in this work. This process was performed using conducting actual data analysis and sensitivity analysis of related coal seam physical properties on four horizontal wells drilled in Upper Freeport coal seam. Results of history matching were compiled to show the initial and existing condition in the coal seam. CO_2 sequestration capacity prediction was completed according to the Langmuir isotherm properties obtained from the history matched reservoir model.

The simulation of CH_4 gasification and CO_2 injection process was quite complicated. The special swelling and shrinkage features and the nature of gas retention in CBM reservoirs make the modeling and history matching of production and injection data in coal bed methane more complex because of the permeability and porosity variations compared to conventional resources.

Sensitivity analysis results suggested that sorption time, cleat permeability, strain, and Langmuir isotherm are the most influential parameters during CH_4 production and CO_2 injection process. It is concluded by the Langmuir isotherm parameters from history matched model that the total CO_2 sequestration capacity is about 22,817 tons excluding the free gas part in the cleat system. The total CO_2 injection amount in the first three years was 4.5×10^7 ft^3 or 2,600 tons, which caused an increase of 6,700 scf/day in CH_4 production rate from other two wells.

ACKNOWLEDGMENTS

This project was funded by the Department of Energy, National Energy Technology Laboratory, Consol Energy, through a support contract with URS Energy and Construction, Inc. The authors want to acknowledge the important contributions of Consol Energy for the field data available for analysis. Acknowledgment is also extended to Tom Wilson in WVU for providing the geological maps for the studied field in this research. Thanks go to Computer Modeling Group Ltd. (CMG) for providing the software to do the PEARL research group at WVU.

REFERENCES

1. S. H. Stevens, D. Spector, and P. Riemer, "Enhanced coalbed methane recovery using CO_2 injection: worldwide resource and CO_2 sequestration potential," in Proceedings of the 6th International Oil & Gas Conference and Exhibition in China (IOGCEC '98), pp. 489–501, Beijing, China, November 1998.
2. J. Ennis-King and L. Paterson, "Engineering aspects of geological sequestration of carbon dioxide," inProceedings of the SPE Asia Pacific Oil and Gas Conference and Exhibition, pp. 134–146, Melbourne, Australia, October 2002.
3. F. M. Orr Jr., "Storage of carbon dioxide in geologic formations," Journal of Petroleum Technology, vol. 56, no. 9, pp. 90–97, 2004.
4. C. Sinayuç and F. Gümrah, "Modeling of ECBM recovery from amasra coalbed in Zonguldak Basin, Turkey," in Proceedings of the Canadian International Petroleum Conference, Alberta, Canada, 2008. View at Publisher · View at Google Scholar
5. R. Petrusak, D. Riestenberg, P. Goad et al., "World class CO_2 sequestration potential in saline formations, oil and gas fields, coal, and shale: the US southeast regional carbon sequestration partnership has it all," inProceedings of the SPE International Conference on CO2 Capture, Storage, and Utilization, pp. 136–153, November 2009.
6. C. L. Liner, "Carbon capture and sequestration: overview and offshore aspects," in Proceedings of the Offshore Technology Conference (OTC '10), pp. 3511–3514, May 2010.

7. J. P. Seidle, "Reservoir engineering aspects of CO_2 sequestration in coals," in Proceedings of the SPE/CERI Gas Technology Symposium, Alberta, Canada, 2000.
8. H. J. M. Pagnier, F. Van Bergen, E. Kreft, L. G. H. Van Der Meer, and H. J. Simmelink, "Field experiment of ECBM-CO_2 in the upper Silesian Basin of Poland (RECOPOL)," in Proceedings of the 67th European Association of Geoscientists and Engineers, EAGE Conference and Exhibition, incorporating SPE (EUROPEC '05), pp. 3013–3015, Madrid, Spain, June 2005.
9. G. A. Hernandez, R. O. Bello, D. A. McVay et al., "Evaluation of the technical and economic feasibility of CO_2 sequestration and enhanced coalbed-methane recovery in Texas low-rank coals," in Proceedings of the SPE Gas Technology Symposium: Mature Fields to New Frontiers, pp. 515–530, Alberta, Canada, May 2006.
10. G. J. Koperna and D. Riestenberg, "Carbon dioxide enhanced coalbed methane and storage: is there promise?" in Proceedings of the SPE International Conference on CO2 Capture, Storage, and Utilization, pp. 183–195, November 2009.
11. J. Q. Shi and S. Durucan, "A model for changes in coalbed permeability during primary and enhanced methane recovery," SPE Reservoir Evaluation and Engineering, vol. 8, no. 4, pp. 291–299, 2005.
12. S. Mazumder and K. H. Wolf, "Differential swelling and permeability change of coal in response to CO_2 injection for ECBM," International Journal of Coal Geology, vol. 74, no. 2, pp. 123–138, 2008. View at Publisher · View at Google Scholar ·
13. L. Dean, "Reservoir engineering for geologists: coalbed methane fundamentals," Reservoir Issue. 2007, 11.
14. Storing CO_2 in Unminable Coal Seams, IEA Greenhouse Gas R&D Programme.
15. K. Aminian and S. Ameri, "Predicting production performance of CBM reservoirs," Journal of Natural Gas Science and Engineering, vol. 1, no. 1-2, pp. 25–30, 2009. View at Publisher · View at Google Scholar ·
16. I. Zulkamain, Simulation study of the effect of well spacing, permeability, anisotropy, Palmar and Mansoori model on coalbed methane production. [M.S. thesis], Texas A&M University, 2005.

17. I. Palmer and J. Mansoori, "How permeability depends on stress and pore pressure in coalbeds: a new model," SPE Reservoir Engineering, vol. 1, no. 6, pp. 539–543, 1998.
18. "CO_2 storage with ECBM study begins in West Virginia," http://www.carboncapturejournal.com/displaynews.php?NewsID=442.
19. D. J. Remner, T. Ertekin, W. Sung, and G. R. King, "Parametric study of the effects of coal seam properties on gas drainage efficiency," SPE Reservoir Engineering, vol. 1, no. 6, pp. 633–646, 1986.
20. A. N. Okeke, Sensitivity analysis of modeling parameters that affect the dual peaking behavior in coalbed methane reservoirs [M.S. thesis], Texas A&M University, 2005.
21. Q. P. Huy, K. Sasaki, Y. Sugai, et al., "Numerical simulation of CO_2 enhanced coal bed methane recovery for A vietmese coal seam," JournaL of NoveL Carbon Resource Sciences, vol. 2, pp. 1–7, 2010.
22. D. Jasinge and P. G. Ranjith, "Carbon dioxide sequestration in geologic formation with special reference to sequestration in deep coal seams," in Proceedings of the 45th U.S. Rock Mechanics/Geomechanics Symposium, 2011.

Chapter 6

Advances in Pressure Swing Adsorption for Gas Separation

Carlos A. Grande

Department of Process Chemistry, SINTEF Materials and Chemistry, Blindern, 0314 Oslo, Norway

ABSTRACT

Pressure swing adsorption (PSA) is a well-established gas separation technique in air separation, gas drying, and hydrogen purification separation. Recently, PSA technology has been applied in other areas like methane purification from natural and biogas and has a tremendous potential to expand its utilization. It is known that the adsorbent material employed in a PSA process is extremely important in defining its properties, but it has also been demonstrated that process engineering can improve the performance of PSA units significantly. This paper aims to provide an overview of the fundamentals of PSA process while focusing specifically on different innovative engineering approaches that contributed to continuous improvement of PSA performance.

INTRODUCTION

Adsorption is the name of the spontaneous phenomenon of attraction that a molecule from a fluid phase experiences when it is close to the surface of a solid, named adsorbent. There are several pristine works that explain this phenomenon in detail [1–18]. Adsorbents are porous solids, preferably having a large surface area per unit mass. Since different molecules have different interactions with the surface of the adsorbent, it is eventually possible to separate them. When the adsorbent is put in contact with a fluid phase, an equilibrium state is achieved after a certain time. This equilibrium establishes the thermodynamic limit of the adsorbent loading for a given fluid phase composition, temperature, and pressure [3]. Information about the adsorption equilibrium of the different species is vital to design and model adsorption processes [19–27]. The time required to achieve the equilibrium state may be also important, particularly when the size of the pores of the adsorbent are close to the size of the molecules to be separated [28–43].

In an adsorption process, the adsorbent used is normally shaped into spherical pellets or extruded. Alternatively, it can be shaped into honeycomb monolithic structures resulting in reduced pressure drop of the system [44–54]. The feed stream is put into contact with the adsorbent that is normally packed in fixed beds. The less adsorbed (light) component will break through the column faster than the other(s). In order to achieve separation, before the other (heavy) component(s) breaks through the column, the feed should be stopped and the adsorbent should be regenerated by desorbing the heavy compound. Since the adsorption equilibrium is given by specific operating conditions (composition, T and P), by changing one of these process parameters it is possible to regenerate the adsorbent.

When the regeneration of the adsorbent is performed by reducing the total pressure of the system, the process is termed pressure swing adsorption (PSA), the total pressure of the system "swings" between high pressure in feed and low pressure in regeneration [55, 56]. The concept was patented in 1932, but its first application was presented thirty years later [57].

Over the years it has been demonstrated that PSA technology can be used in a large variety of applications: hydrogen purification [58–72], air separation [57, 73–80], OBOGS (on-board gas generation system) [81], CO_2 removal [82–84], noble gases (He, Xe, Ar) purification [85–87], CH_4 upgrading [31, 34, 37, 40, 42,88–96], n-iso paraffin separation [5, 97–99], and so forth. The PSA processes are normally associated to low energy consumption when compared to other technologies [12, 55, 100–102].

As a rule of thumb, pressure swing adsorption is preferred to other processes when the concentration of the components to be removed is quite important (more than a few per cent). In such conditions, loading the column with the heavy component is accomplished quite fast and since the pressure of the system can be changed rapidly, the time between adsorption and regeneration is balanced. When the concentration is low, the adsorption step may take much longer and other options like temperature swing adsorption (TSA) can be considered [12].

The behaviour of the PSA unit is mainly determined by the adsorbent employed for the separation. However, the engineering of the PSA unit is also an important aspect. In fact, the main task of defining a PSA unit is to select correctly the adsorbent to be employed [103]. After that, all the engineering efforts should be placed in defining an effective strategy to regenerate the adsorbent. Thus, the advances obtained in PSA units can be divided in two main domains: the discovery of new adsorbents (material science) and new and more efficient ways to use and regenerate the adsorbent (engineering).

This work provides an overview of PSA processes and its evolution on time. The most important industrial applications of PSA processes will be used to address its technological evolution: air separation and hydrogen purification. A growing market of PSA, CH_4-CO_2 separation, will also be used for some specific examples. Although it is not intended to describe the state-of-the-art of materials science, an example of the effect of different adsorbent materials in PSA operation will be provided. Finally, the effect of different regeneration protocols and the reduction of the overall cycle time (Rapid Pressure Swing Adsorption) are discussed.

FUNDAMENTALS OF PRESSURE SWING ADSORPTION

The essential feature of the PSA is that when the adsorbent is saturated, using a sequential valve arrangement, the feed is stopped and simultaneously the total pressure of the column is reduced. The reduction in pressure results in a partial desorption of all the species loaded in the column, "regenerating" the adsorbent. Since this process was patented after TSA, it was initially known as "heatless" process. The first patent application where PSA technology was described, was presented by Charles Skarstrom for oxygen enrichment [57]. A scheme of the two-column PSA introduced in that patent is shown in Figure 1. In order to operate such unit cyclically, a column experiences a series of "steps": events like opening and closing valves and changing flowrate direction for example. The sum of all the steps is termed as "cycle". Even when the process is unsteady, after some cycles it reaches a Cyclic Steady State, CSS. When CSS is achieved, the performance of the cycles of the PSA is constant over time. It should be noted that since this process sometimes involve substantial amount of heat generation, there can be multiple CSS [104].

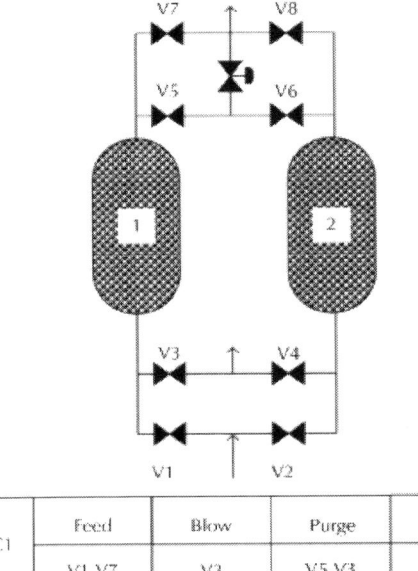

Figure 1: Schematic design of the first two-column pressure swing adsorption unit and valve sequencing for different steps in the cycle [1].

The four steps of the "Skarstrom cycle" are also shown in Figure 1: feed, blowdown (or evacuation), purge and pressurization. In this cycle, in the feed step, air is fed to the first column (C1) at a pressure higher than atmospheric. The adsorbent initially used (zeolite 5A) is selective to nitrogen, making the exiting stream (after valve V7) richer in oxygen. When the adsorbent packed in C1 is saturated and cannot adsorb more nitrogen, the feed is directed to the second column (C2). In order to release part of the nitrogen adsorbed in C1, the flow direction is reversed and the total pressure of the column is reduced by venting to atmosphere (opening valve V3). There are different terms to call this step, but blowdown is one of the most common and will be used here. In the blowdown step, nitrogen is desorbed from the adsorbent and released and at the end of this step, the gas phase inside the column is rich in nitrogen. To additionally remove nitrogen from the column, a purge step (or light gas recycle) is used. The purge consists of recycled part of the enriched air from the other column which is flowing by the pressure differential between the two columns. After the adsorbent is ready to load more nitrogen, the overall pressure of the system should be restored. That is done in the pressurization step using the feed stream. After all these steps were finished, a complete cycle was completed. It is important to notice that although the column operation is discontinuous, the feed stream is being employed so the process can be viewed as continuous. However, the exit is discontinuous and a tank is required to be coupled for a continuous discharge. Also, the operation in both columns should be synchronized to satisfy the continuous utilization of the feed and to provide purge gas to the other column.

The requirement of continuous feed processing, even being a discontinuous process, was recognized, since one of the first inventions of adsorption processes [105]. Furthermore, the valve arrangement for sequential opening - close and step definition was also very similar to designs presented for TSA processes [106]. However, the contribution of Skarstrom allowed a tremendous improvement in utilization of the adsorbents: while TSA cycles last for several hours, the PSA cycles are much shorter and thus using more adsorbent per unit time.

Another important aspect of a PSA process was mentioned in Skarstrom's application: heat effects and conservation. In the adsorption step, heat generated by adsorption may be important in which case the temperature of the column changes with time and also with position [4, 5, 55]. The consequence is a reduction in the adsorbent capacity. The "heat effects" may be very important in designing a PSA unit [107] and should be taken into account in the design: laboratory or small-scale experiments are either isothermal or close to isothermal and the heat capacity of the wall is important while large-scale processes behave adiabatically. In the desorption steps, the opposite is happening: energy is required for desorption resulting in a temperature decrease enhancing the potential capacity of the adsorbent and making desorption more difficult. This will happen in all PSA applications but in some cases, the amount of heat generated is not so important and the process can be considered isothermal. Every time there is a temperature swing associated to the PSA cycle, the performance is worse than what would be if the cycle is isothermal. However, since the thermal effects are present, it is good practice to conserve the "heat wave" inside the column: this heat will be used for a faster desorption.

MODIFICATIONS TO THE SKARSTROM CYCLE: NEW CYCLE STEPS

In the years after Skarstrom invention, there were several patent applications to improve the cycle. In a patent that was filled almost at the same time as Skarstrom, the regeneration under vacuum was introduced by Guerin de Montgareuil and Domine [73]. When vacuum is used for regeneration it is common to term the unit as vacuum pressure swing adsorption (VPSA). Although the utilization of vacuum may have an impact on the energetic requirements of the system, the efficiency of the unit may be greatly improved if the loading of the most adsorbed components changes dramatically at pressure lower than atmospheric. In the same invention, the authors have introduced the utilization of the pressurization step using part of the enriched gas. The utilization of a pressurization using part of the purified gas had impact in the purity of the produced gas [108]. Even when using the same pressure swing concept, the alternatives to develop the PSA technology are quite diverse, opening the "PSA engineering" possibilities.

The introduction of a pressure equalization step was developed at ESSO Research group [74, 109, 110]. Taking the two-column PSA scheme from Figure 1, after C1 ends the feed step (and is at high pressure), C2 ends the purge step (and is at low pressure). In that moment, V5 and V6 are simultaneously opened, short-circuiting the columns. This means that part of the gas that will normally get lost in the blowdown step is being used to pressurize the other column, loosing less purified gas. If the gas moving from one column to the other is not significantly adsorbed (e.g., hydrogen) the pressure achieved after the equalization step is the geometric average between these two values. The overall pressure can be lower if the gas transferred is fast adsorbed [111]. The result of the pressure equalization step is a direct improvement in the recovery of the light product [112, 113]. The introduction of a pressure equalization step in a 2-column PSA unit results in a significant change of the "continuity" of the process. When the two columns are in pressure equalization, there is no feed processing so at least one more column is required [110].

When several columns are employed, several pressure equalization steps can be done [114–116] and as a consequence, the overall recovery is increased [65, 117, 118]. This finding resulted in the design of multiple column (Polybed) PSA units [65].

Another possibility to remove part of the light component from the column before blowdown is depressurizing the bed co-currently to the feed direction. This step is very useful in hydrogen purification and is normally termed as "provide purge" step since it provides gas for purging other column [119].

Co-n-current depressurization was also used to remove the less adsorbed gas from the column in order to increase the content of most-adsorbed gas inside the column (aiming to its concentration) [32, 120–122].

An interesting concept of column depressurization is provided by the unique availability of "free vacuum" obtained in outer space [123]. In order to have a faster depressurization, it was proposed to open the column from both ends to release the gas faster. Parallel equalization using valves at different column lengths were also suggested [124]. Using a low-pressure feed as a purge was also suggested to increase the purity and recovery when compared with the Skarstrom cycle [125]. For the case of separation of a ternary mixture, feeding and one product withdrawal in intermediate positions of the column was also suggested, with a PSA design resembling the Petliuk scheme for distillation [126, 127].

In order to displace the light component to the product end, a recycle of the heavy component was suggested by Basmadjian and Pogorski [128]. This step was called "rinse." Although the rinse step aimed to provide a solution to concentration of low-per cent light compounds, it has been widely used for other purposes: concentration of the more adsorbed species [32, 120–122, 129–132].

In fact, the number of possible "steps" is not very large. However, using them in an efficient way has proved to be a difficult task. So far, the question raised by Professor Ruthven in 1992 was not yet completely answered [133] ("Is it possible to develop an algorithm for automatic generation of PSA cycles and tuning of the various steps?").

PERFORMANCE INDICATOR PARAMETERS OF A PSA PROCESS

So far, it has been shown that PSA processes have a tremendous flexibility in design (so large that sometimes is misleading). A completely different number of columns can be used and also a quite large number of cycles are possible. In order to provide a certain "common framework" to understand some aspects of PSA engineering, it is desirable to have some "performance indicators" (PI) which will be the ones that will define how well performs the PSA process. For the definition of such parameters, the PSA process depicted in Figure2 can be considered. The image shows a PSA process with x columns (x can also be unity) accommodating a specific mass of adsorbent per column (w_{ads}) and with multiple connection lines to accommodate very different steps. The objective is to separate component i from N components and two cases may be found: either the purpose of the PSA is to purify the less adsorbed gas or alternatively, to concentrate the more adsorbed gas.

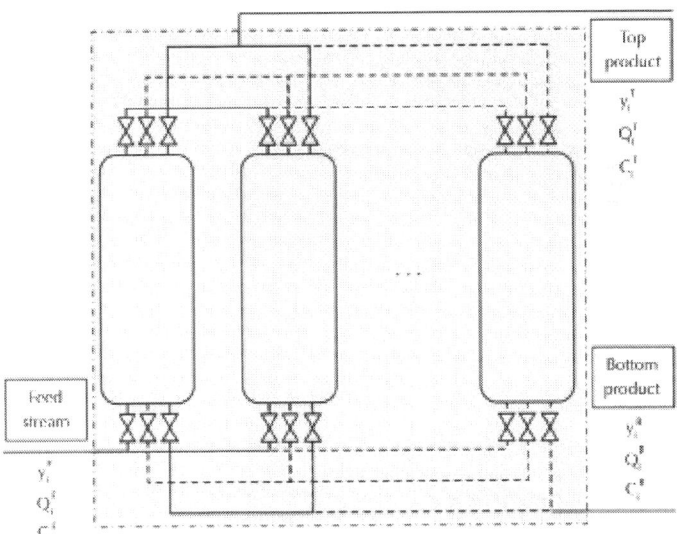

Figure 2: "Grey-box" generic example of a pressure swing adsorption (PSA) process. The inlet and exit streams are characterized by molar fraction (y_i), volumetric flowrate (Q_i, m³/s), and gas concentration (C_i, mol/m³).

The most common PI found in PSA processes are listed in Table 1 [134]. The two first PI (purity and recovery) are related to the separation efficiency of the PSA and normally establish the GO/NO GO condition in process design. If such specifications are satisfied, the "fingerprint" of the unit is evaluated by the productivity. Finally, the energetic considerations should be made. Since the process is so flexible, it is difficult to define an energetic PI other than saying that is the sum of all work used for compression and vacuum. Note that the recovery and productivity have an integral term that is mainly due to variations in flowrate in the exit streams.

Table 1: Performance indicator parameters for a PSA process

Less-adsorbed gas is the product	More-adsorbed gas is the product
Purity $= \dfrac{C_i^T}{\sum_{i=1}^{N} C_i^T} = y_i^T$	Purity $= \dfrac{C_i^B}{\sum_{i=1}^{N} C_i^B} = y_i^B$
Recovery $= \dfrac{\int_0^t C_i^T Q_i^T dt}{C_i^F Q_i^F}$	Recovery $= \dfrac{\int_0^t C_i^B Q_i^B dt}{C_i^F Q_i^F}$
Productivity $= \dfrac{\int_0^{t_{cycle}} C_i^T Q_i^T dt}{X w_{ads} t_{cycle}}$	Productivity $= \dfrac{\int_0^{t_{cycle}} C_i^B Q_i^B dt}{X w_{ads} t_{cycle}}$
Energy = sum of all compression and vacuum sources used	

Most works on PSA processes have shown that normally the purity and recovery present a trade-off for the design. In the case of recovering the less adsorbed gas, if more purge is used, more of the contaminants can be desorbed from the column and purity increases, but since more light gas is exiting from the "bottom end," light-gas recovery is smaller. A similar effect is observed for the utilization of the rinse step and purity and recovery of the more adsorbed gas.

However, other strategies are valid to improve process recovery without seriously affecting the purity. The case of Polybed PSA for H_2 purification is a good example [65]. The units built until 1975 were having 4 columns and the recovery of H_2 was around 60%. Nowadays, PSA unit with 12 columns are found [65] and up to 16 columns were patented [135] with H_2 recovery close to 90%. When the number of columns is increased, more pressure equalization steps can be performed and thus less hydrogen is lost with the contaminants, increasing its recovery.

The developments in the PSA process presented above were mainly motivated to improve the purity and the recovery of the target product(s). Nowadays, several new applications of PSA as an alternative technology are still in the stage of finding proper cycle configuration (step scheduling and times, number of columns, etc.). Other applications in more established markets are intending to improve either the unit size and/or the energetic consumption of the separation.

THE ROLE OF THE ADSORBENT IN PSA

The development of materials science in the last 60 years was quite intense. The result was the discovery of many porous materials, from all kind of zeolites and mesoporous materials [136–141] to the most diverse surfaces in activated carbons [142–145] and lately the high-surface area coordination polymers [146–151]. However, as strange as it may seem, only few materials are used in PSA units nowadays.

A review of adsorption properties of the different materials is out of the scope of this work, but good databases with adsorption properties of different gases on several adsorbents can be found [16, 152, 153]. What is important to mention is that a material to be used in PSA should be easily regenerated. It is frequent to find in literature adsorbents with a very high capacity, particularly at low pressures. Normally the isotherms of gases on such adsorbents are "rectangular": very steep at low pressures and quite flat after a certain pressure. Defining the "cyclic capacity" as the difference of loading between the high and low pressures of the PSA cycle, the only way to have an acceptable cyclic capacity is making blowdown at very high vacuum. The direct implication of using such conditions is that the power consumption increases rapidly. So, materials showing linear or slightly nonlinear isotherms are preferred in PSA design.

One frequent case is to have a multicomponent mixture of gases and that the number of compounds to be separated cannot be removed by a single adsorbent. The solution to this problem was found for the case of H_2 purification from methane steam reforming. In this application, H_2 is mixed with H_2O, CO_2, CO, unconverted CH_4, and possibly other gases like N_2. Activated carbon can be used to remove H_2O and CO_2 quite selectively but the loading for CO is rather limited for small partial pressures. It is thus common practice to use different layers of adsorbents to increase the loading of CO in the same column. This approach has also been applied in other separations [66, 70, 79, 154–160]. Consecutive layers of adsorbents can also be used to improve the productivity of kinetic adsorbents by adding a material that can be easily regenerated after the kinetic adsorbent [161, 162].

Other important aspect regarding the material properties for PSA applications is the diffusion of the different gases through its porous structure. There are different types of "resistances" to diffuse from the bulk gas phase to the adsorption site [4, 5]. They are: boundary layer around the adsorbent particle, and resistances in the macro-meso pores, mouth of the micropores, and micropores (or crystals).

In some applications however, these mass transfer "problems" have become part of the solution. In fact, if the diffusional resistance of one of the components of the mixture is very large, this gas will take so long to adsorb that can be separated from other gas that diffuses faster through the pores.

The "kinetic processes" were recognized soon [28]. In fact, materials like zeolites are called "molecular sieves" because of this effect [136]. Another example of kinetic materials is the carbon molecular sieves (CMS) [29–31, 33, 38, 163–167]. A CMS is prepared by contracting the pores of an activated carbon to limit the adsorption of some molecules. Its first utilization was for air separation to separate O_2 from N_2.

An extreme example of resistance to diffusion is the molecular exclusion like in the Isosiv process [5, 97–99]. In the Isosiv process, n-paraffins are selectively adsorbed in zeolite 5A, while isoparaffins are kinetically excluded from the zeolite crystals.

Most recently, several inorganic materials have proved to be useful for kinetic separations [34, 36, 168–173]. A special kind of titanosilicates, ETS-4, cation exchanged with alkali-earth metals can be used for kinetic separations [35, 41, 174, 175]. In these materials, the pore size can be tuned with a very high accuracy by thermal treatment of the sample. Many studies have confirmed that CH_4 can be excluded from the structure while gases like H_2S, CO_2, and specially N_2 can be adsorbed [43, 176, 177].

ADVANCES IN PROCESS ENGINEERING

From all the main advances in process engineering, the most challenging one is the development of cyclic strategies that can improve the performance indicators of the PSA. Despite the performance of the material, the design of a PSA process requires several engineering decisions that should be taken sometimes with a very deep impact in terms of performance indicators. The main drawback of the engineering of a PSA process is that it is quite task consuming (and normally iterative).

With modern computers, the design of the PSA cycle can be carried out by modelling different scenarios. There are different degrees of complexity to define a PSA model, normally comprising several partial differential equations linked by the equation of state and the isotherm model to define the thermodynamic properties of the gas and adsorbed phases, respectively. Although the model can be solved by numerical methods [55, 113, 178–183], there are several commercial programs that can be already used for that purpose: ASPEN, COMSOL, gPROMS, PROSIM, and so forth [18, 184–187].

The simulation of a PSA process requires an initial step of defining a cycle structure (ordering the steps in a pre-defined sequence) and then estimate the performance indicators obtained. For the selected cycle, all the step times, blowdown pressure, and flowrates of rinse and purge steps should be determined [25, 188–192]. Alternatively, it has been suggested that a general "super-cycle" can be used to estimate the optimal duration of each of the steps [193].

In most cases, the definition of the cycle has to be done under certain constrains like combining it in a multiple column array. Other constraints can result from the availability of gas to the purge step, the continuous utilization of vacuum pump for blowdown, and so forth. The availability of gas to the purge step can also proceed from a depressurization step (provide purge) [119] or from a prestored amount in a tank [194]. A graphical procedure to schedule PSA cycles was suggested [195, 196]. It is also found in literature that in some cases, the best cycle does not match perfectly in a continuous array of columns and thus an "idle" step is used where the column is closed and no effective step for adsorption or desorption takes place. However, the existence of idle periods does result in smaller unit productivity of the PSA unit.

Recasting how the PSA productivity is calculated, we can see the interaction between the influence of process engineering and adsorbent development is mixed. If we have an adsorbent with a better cyclic capacity, we will be able to adsorb more gas per cycle and thus reduce the overall weight of adsorbent (or alternatively, increase the production of gas). On the other side, by better process engineering, we could improve the performance of the unit by balancing the amount of gas produced and possibly reducing the number of columns employed.

Furthermore, there is a third alternative: reduce the total cycle time. This alternative was suggested many years ago [197] and has started been implemented in the 80s [198]. When the total cycle time is smaller than 30 seconds, the process is normally called Rapid PSA (RPSA) [145, 179, 198–214].

A typical cycle time (t_{cycle}) of a normal PSA process is in the order of 10 minutes. In that time, the adsorbent is used to adsorb and desorb a certain amount of gas. Within each column of the PSA that amount adsorbed will be distributed in an initial zone where equilibrium has been achieved and a "mass transfer" zone close to the end of the column where the adsorbent is not completely saturated. The mass transfer zone is related to kinetic limitations to diffuse into the adsorbent and axial dispersion. Reducing the cycle time will result in more kinetic limitations and thus longer mass transfer zones. However, if reducing the cycle time in a factor of 10 results in a decrease of the amount adsorbed/desorbed in a factor of 2 (by kinetic limitations to adsorb), then the overall productivity of the PSA unit has still increased in a factor of 5. The result is that the PSA unit will be five times smaller!

There are several fields where RPSA can make a complete difference. A PSA for production of medicinal oxygen is a very suitable unit for utilization in hospitals. However, the concept of RPSA has opened the possibility of portable devices with quite small size that can be used for ambulatory patients with chronic lung diseases [78, 215]. Comparing the productivity of a PSA process to purify hydrogen, it can be noted that is quite lower than the productivity found in other

PSA applications. In such a field, the utilization of RPSA concept can lead to significant reduction in size [201, 216].

The utilization of RPSA is limited by fluid dynamics. Using the Ultra-rapid piston driven PSA, the total cycle time was less than 5 seconds (its adsorption/desorption cycles resemble the expansion and compression of an internal combustion engine). Under such conditions, the mathematical models used to simulate normal PSA processes may not work [210, 217]: mass and energy transfer description using simplifications like LDF (linear driving force) are not applicable. There are also some particularities related to RPSA that could be overcome with the utilization of specialized devices.

In RPSA processes, the time required for pressurization of the bed can be a problem. It has been proven that by using a honeycomb monolith, it is possible to reduce the pressure drop of the PSA process [209] and thus reduce the overall pressurization time. Alternative to monolithic structures, laminated adsorbents have been suggested [218].

The other invention that is directly applicable to RPSA technology is the rotary valve [205, 207, 219]. Taking as example the PSA unit shown in Figure 1, it can be observed that the step changes in a normal PSA are accomplished by the simultaneous operation of a sometimes complex valve array. Using rotary multiport valves, it is possible to change the events taking place in all the columns at the exact same time. Using a normal valve array, a failure of one second in opening or closing one of the valves can have a significant impact in a RPSA cycle.

Another approach to PSA technology was carried out using radial columns [220–222]. Using radial columns, the length of adsorbent is normally small (resulting in decreased pressure drop) and the amount of gas to be treated at a reasonable gas velocity can be higher.

CONCLUDING REMARKS

The great flexibility of PSA is normally associated to process complexity and is still one of the major issues to introduce this technology in several fields of industry. On the other hand, the large flexibility of PSA processes still constitutes its main advantage and may be the reason why it has found applications in diverse fields.

PSA technology can be considered a mature technology in air separation, drying, and hydrogen purification, but there is plenty of work to do to establish this technique in other fields [223]. Many researchers around the world are currently working on CO_2 capture from flue gases. It has been potentially demonstrated that CO_2 can be captured using PSA [224–227] but more fundamental and long-term pilot plant studies are required to properly benchmark this technique against amines. Also, olefin-paraffin separation by adsorption was quite studied, but the energetic consumption of the separation by adsorption is still comparable to distillation [228]. Utilization of PSA for natural gas upgrading (CH_4-CO_2 separation basically) still also remains a challenge [229, 230]. PSA technology and even RPSA can be used to upgrade biogas, but the flowrate and pressure levels of natural gas require alternative solutions. Furthermore, new stringent legislation related to reducing the emission of greenhouse gases is changing the design of processes in energy and fuel industries. New processes intend to include or integrate the CO_2 capture, thus introducing specifications in the most adsorbed compound. A solution that is already in use and should be more explored is the dual PSA concept [231–235].

In all these emergent applications of PSA technology, faster and better solutions can happen by having a good interaction between materials science and process engineering.

REFERENCES

1. I. Langmuir, "The adsorption of gases on plane surfaces of glass, mica and platinum," The Journal of the American Chemical Society, vol. 40, no. 9, pp. 1361–1403, 1918.
2. M. Polanyi, "Section III.—theories of the adsorption of gases: a general survey and some additional remarks. Introductory paper to section III," Transactions of the Faraday Society, vol. 28, pp. 316–333, 1932.
3. A. L. Myers and J. M. Prausnitz, "Thermodynamics of mixed-gas adsorption," AIChE Journal, vol. 11, pp. 121–127, 1965.
4. D. M. Ruthven, Priciples of Adsorption and Adsorption Processes, John Wiley & Sons, New York, NY, USA, 1984.
5. R. T. Yang, Adsorbents. Fundamentals and Applications, John Wiley & Sons, New Jersey, NJ, USA, 2003.

6. P. C. Wankat, Large-Scale Adsorption and Chromatography, CRC Press, Boca Raton, Fla, USA, 1986.
7. A. E. Rodrigues, M. D. LeVan, and D. Tondeur, Adsorption, Science and Technology, Kluwer Academic Publishers, Boston, Mass, USA, 1989.
8. M. Suzuki, Adsorption Engineering, Chemical Engineering Monographs, Elsevier, Tokyo, Japan, 1990.
9. J. Kärger and D. M. Ruthven, Diffusion in Zeolites and Other Microporous Solids, John Wiley & Sons, London, UK, 1992.
10. C. Tien, Adsorption Calculations and Modeling, Butterworth-Heinemann, Boston, Mass, USA, 1994.
11. D. Basmadjian, The Little Adsorption Book: A Practical Guide for Engineers and Scientists, CRC Press, Boca Raton, Fla, USA, 1997.
12. J. L. Humphrey and G. E. Keller, Separation Process Technology, McGraw-Hill, New York, NY, USA, 1997.
13. D. D. Do, Adsorption Analysis: Equilibria and Kinetics, Imperial College Press, London, UK, 1998.
14. J. W. Thomas and B. D. Crittenden, Adsorption Technology and Design, Elsevier, Boston, Mass, USA, 1998.
15. O. Talu, "Needs, status, techniques and problems with binary gas adsorption experiments," Advances in Colloid and Interface Science, vol. 76-77, pp. 227–269, 1998.
16. F. Rouquerol, J. Rouquerol, and K. Song, Adsorption by Powders and Porous Solids, Academic Press, London, UK, 1999.
17. J. Keller and R. Staudt, Gas Adsorption Equilibria: Experimental Methods and Adsorption Isotherms, Springer, Boston, Mass, USA, 2005.
18. P. C. Wankat, Separation Process Engineering, Prentice Hall, London, UK, 2nd edition, 2007.
19. K. S. Knaebel and F. B. Hill, "Pressure swing adsorption: development of an equilibrium theory for gas separations," Chemical Engineering Science, vol. 40, no. 12, pp. 2351–2360, 1985.
20. M. D. LeVan, "Pressure swing adsorption: equilibrium theory for purification and enrichment," Industrial and Engineering Chemistry Research, vol. 34, no. 8, pp. 2655–2660, 1995.

21. G. Pigorini and M. D. LeVan, "Equilibrium theory for pressure swing adsorption. 2: purification and enrichment in layered beds," Industrial and Engineering Chemistry Research, vol. 36, no. 6, pp. 2296–2305, 1997.
22. G. Pigorini and M. D. LeVan, "Equilibrium theory for pressure swing adsorption. 3: separation and purification in two-component adsorption," Industrial and Engineering Chemistry Research, vol. 36, no. 6, pp. 2306–2319, 1997.
23. G. Pigorini and M. D. LeVan, "Equilibrium theory for pressure-swing adsorption. 4: optimizations for trace separation and purification in two-component adsorption," Industrial and Engineering Chemistry Research, vol. 37, no. 6, pp. 2516–2528, 1998.
24. A. Serbezov and S. V. Sotirchos, "Semianalytical solution for multicomponent pressure swing adsorption," Chemical Engineering Science, vol. 53, no. 20, pp. 3521–3536, 1998.
25. A. Serbezov, "Effect of the process parameters on the lenght of the mass transfer zone during product withdrawal in pressure swing adsorption cycles," Chemical Engineering Science, vol. 56, no. 15, pp. 4673–4684, 2001.
26. A. D. Ebner and J. A. Ritter, "Equilibrium theory analysis of rectifying PSA for heavy component production," AIChE Journal, vol. 48, no. 8, pp. 1679–1691, 2002.
27. A. D. Ebner and J. A. Ritter, "Equilibrium theory analysis of dual reflux PSA for separation of a binary mixture," AIChE Journal, vol. 50, no. 10, pp. 2418–2429, 2004.
28. H. W. Habgood, "The kinetics of molecular sieve action: sorption of nitrogen-methane mixtures by Linde Molecular Sieve 4A," Canadian Journal of Chemistry, vol. 36, pp. 1384–1397, 1958.
29. K. Chihara, M. Suzuki, and K. Kawazoe, "Adsorption rate on molecular sieving carbon by chromatography," AIChE Journal, vol. 24, no. 2, pp. 237–246, 1978.
30. H. Jüntgen, K. Knoblauch, and K. Harder, "Carbon molecular sieves: production from coal and application in gas separation," Fuel, vol. 60, no. 9, pp. 817–822, 1981.
31. A. Kapoor and R. T. Yang, "Kinetic separation of methane-carbon dioxide mixture by adsorption on molecular sieve carbon," Chemical Engineering Science, vol. 44, no. 8, pp. 1723–1733, 1989.

32. R. Ramachandran, L. H. Dao, and B. Brooks, "Method of producing unsaturated hydrocarbons and separating the same from saturated hydrocarbons," U.S. patent 5, 365, 011, 1994.
33. A. I. Fatehi, K. F. Loughlin, and M. M. Hassan, "Separation of methane-nitrogen mixtures by pressure swing adsorption using a carbon molecular sieve," Gas Separation and Purification, vol. 9, no. 3, pp. 199–204, 1995.
34. M. W. Seery, "Bulk separation of carbon dioxide from methane using natural clinoptilolite," World Patent, WO 98/58726, 1998.
35. S. M. Kuznicki, V. A. Bell, I. Petrovic, and P. W. Blosser, "Separation of nitrogen from mixtures thereof with methane utilizing barium exchanged ETS-4," US patent no. 5, 989, 316, 1999.
36. J. Padin, S. U. Rege, R. T. Yang, and L. S. Cheng, "Molecular sieve sorbents for kinetic separation of propane/propylene," Chemical Engineering Science, vol. 55, no. 20, pp. 4525–4535, 2000.
37. M. Mitariten, "New technology improves nitrogen-removal economics," Oil and Gas Journal, vol. 99, no. 17, pp. 42–44, 2001.
38. A. Jayaraman, A. S. Chiao, J. Padin, R. T. Yang, and C. L. Munson, "Kinetic separation of methane/carbon dioxide by molecular sieve carbons," Separation Science and Technology, vol. 37, no. 11, pp. 2505–2528, 2002.
39. W. B. Dolan and M. J. Mitariten, "Heavy hydrocarbon recovery from pressure swing adsorption unit tail gas," 2003, US patent 6, 610, 124.
40. W. B. Dolan and M. J. Mitariten, "CO_2 rejection from natural gas," US patent 2003/0047071, 2003.
41. S. M. Kuznicki and V. A. Bell, "Olefin separation employing ETS molecular sieves," U.S. patent, 6, 517, 611, 2003.
42. M. B. Kim, Y. S. Bae, D. K. Choi, and C. H. Lee, "Kinetic separation of landfill gas by a two-bed pressure swing adsorption process packed with carbon molecular sieve: nonisothermal operation," Industrial and Engineering Chemistry Research, vol. 45, no. 14, pp. 5050–5058, 2006.
43. S. Cavenati, C. A. Grande, F. V. S. Lopes, and A. E. Rodrigues, "Adsorption of small molecules on alkali-earth modified titanosilicates," Microporous and Mesoporous Materials, vol. 121, no. 1–3, pp. 114–120, 2009.

44. D. B. Shah, S. P. Perera, and B. D. Crittenden, "Adsorption dynamics in a monolithic adsorbent," in Fundamentals of Adsorption, M. D. LeVan, Ed., Kluwer Academic Publishers, Boston, Mass, USA, 1996.
45. K. P. Gadkaree, "System and method for adsorbing contaminants and regenerating the adsorber," U.S. patent 5, 658, 372, 1997.
46. Y. Y. Li, S. P. Perera, and B. D. Crittenden, "Zeolite monoliths for air separation—part 2: oxygen enrichment, pressure drop and pressurization," Chemical Engineering Research and Design, vol. 76, no. 8, pp. 931–941, 1998. · ·
47. R. Jain, A. I. LaCava, A. Maheshwary, J. R. Ambriano, D. R. Acharya, and F. R. Fitch, "Air separation using monolith adsorbent bed," U. S. patent, 6, 231, 644, 2001.
48. R. E. Critoph, "Multiple bed regenerative adsorption cycle using the monolithic carbon-ammonia pair," Applied Thermal Engineering, vol. 22, no. 6, pp. 667–677, 2002.
49. D. J. Kim, J. W. Kim, J. E. Yie, and H. Moon, "Temperature-programmed adsorption and characteristics of honeycomb hydrocarbon adsorbers," Industrial and Engineering Chemistry Research, vol. 41, no. 25, pp. 6589–6592, 2002.
50. T. Valdés-Solís, M. J. G. Linders, F. Kapteijn, G. Marbán, and A. B. Fuertes, "Adsorption and breakthrough performance of carbon-coated ceramic monoliths at low concentration of n-butane," Chemical Engineering Science, vol. 59, no. 13, pp. 2791–2800, 2004.
51. A. B. Gorbach, M. Stegmaier, G. Eigenberger, J. Hammer, and H. G. Fritz, "Compact pressure swing adsorption processes-impact and potential of new-type adsorbent-polymer monoliths," Adsorption, vol. 11, no. 1, pp. 515–520, 2005. · ·
52. C. A. Grande, S. Cavenati, P. Barcia, J. Hammer, H. G. Fritz, and A. E. Rodrigues, "Adsorption of propane and propylene in zeolite 4A honeycomb monolith," Chemical Engineering Science, vol. 61, no. 10, pp. 3053–3063, 2006. · ·
53. I. Perdana, D. Creaser, I. Made Bendiyasa, Rochmadi, and B. Wikan Tyoso, "Modelling NOx adsorption in a thin NaZSM-5 film supported on a cordierite monolith," Chemical Engineering Science, vol. 62, no. 15, pp. 3882–3893, 2007.

54. F. Rezaei and P. Webley, "Optimum structured adsorbents for gas separation processes," Chemical Engineering Science, vol. 64, no. 24, pp. 5182–5191, 2009. ·
55. D. M. Ruthven, S. Farooq, and K. S. Knaebel, Pressure Swing Adsorption, VCH Publishers, New York, NY, USA, 1994.
56. D. Tondeur and P. C. Wankat, "Gas purification by PSA," Separation and Purification Methods, vol. 14, no. 2, pp. 157–212, 1985.
57. C. W. Skarstrom, "Method and apparatus for fractionating gas mixtures by adsorption," U.S. patent 2, 944, 627, 1960.
58. S. Sircar, "Separation of multicomponent gas mixtures," U.S. patent 4, 171, 206, 1979.
59. P. Cen and R. T. Yang, "Separation of a five-component gas mixture by pressure swing adsorption," Separation Science and Technology, vol. 20, no. 9-10, pp. 725–747, 1985.
60. S. Sircar, "Fractionation of multicomponent gas mixtures by pressure swing adsorption," U.S. patent 4, 790, 858, 1988.
61. T. C. Golden, R. Kumar, and W. C. Kratz, "Hydrogen purification," US patent 4, 957, 514, 1990.
62. R. Kumar, "Adsorption process for recovering two high purity gas products from multicomponent gas mixtures," U.S. patent 4, 913, 709, 1990.
63. O. Bomard, J. Jutard, S. Moreau, and X. Vigor, "Method for purifying hydrogen based gas mixtures using a lithium-exchanged X zeolite," W.O. Patent 97/45363, 1997.
64. A. Malek and S. Farooq, "Hydrogen purification from refinery fuel gas by pressure swing adsorption," AIChE Journal, vol. 44, no. 9, pp. 1985–1992, 1998.
65. J. Stöcker, M. Whysall, and G. Q. Miller, 30 Years of PSA Technology for Hydrogen Purification, UOP LLC, Des Plaines, Ill, USA, 1998.
66. C. H. Lee, J. Yang, and H. Ahn, "Effects of carbon-to-zeolite ratio on layered bed H_2 PSA for coke oven gas," AIChE Journal, vol. 45, no. 3, pp. 535–545, 1999.
67. S. Sircar, W. E. Waldron, M. B. Rao, and M. Anand, "Hydrogen production by hybrid SMR-PSA-SSF membrane system," Separation and Purification Technology, vol. 17, no. 1, pp. 11–20, 1999.

68. J. H. Park, J. N. Kim, and S. H. Cho, "Performance analysis of four-bed H_2 PSA process using layered beds," AIChE Journal, vol. 46, no. 4, pp. 790–802, 2000.
69. S. Sircar and T. C. Golden, "Purification of hydrogen by pressure swing adsorption," Separation Science and Technology, vol. 35, no. 5, pp. 667–687, 2000.
70. J. G. Jee, M. B. Kim, and C. H. Lee, "Adsorption characteristics of hydrogen mixtures in a layered bed: binary, ternary, and five-component mixtures," Industrial and Engineering Chemistry Research, vol. 40, no. 3, pp. 868–878, 2001.
71. M. S. A. Baksh, M. W. Ackley, and F. Notaro, "Process and apparatus for hydrogen purification," W.O. Patent 2004/058630, 2004.
72. R. L. Bec, "Method for purifying hydrogen-based gas mixtures using calcium X-zeolite," U.S. patent 6, 849, 106, 2005.
73. P. Guerin de Montgareuil and D. Domine, "Process for separating a binary gaseous mixture by adsorption," US patent 3, 155, 468, 1964.
74. N. H. Berlin, "Method for providing an oxygen-enriched environment," U.S. Patent 3, 280, 536, 1966.
75. H. Jüntgen, K. Knoblauch, J. Reichenberger, and F. Tarnow, "Process for the recovery of nitrogen-rich gases from gases containing at least oxygen as other component," U.S. patent 4, 264, 339, 1981.
76. D. M. Ruthven, N. S. Raghavan, and M. M. Hassan, "Adsorption and diffusion of nitrogen and oxygen in a carbon molecular sieve," Chemical Engineering Science, vol. 41, no. 5, pp. 1325–1332, 1986.
77. C. G. Coe, J. F. Kirner, R. Pierantozzi, and T. R. White, "Nitrogen adsorption with Ca and or Sr exchanged lithium X-zeolites," U.S. patent 5, 152, 813, 1992.
78. J. G. Jee, J. S. Lee, and C. H. Lee, "Air separation by a small-scale two-bed medical O_2 pressure swing adsorption," Industrial and Engineering Chemistry Research, vol. 40, no. 16, pp. 3647–3658, 2001.
79. Y. Lü, S. J. Doong, and M. Bülow, "Pressure-swing adsorption using layered adsorbent beds with different adsorption properties:

II-experimental investigation," Adsorption, vol. 10, no. 4, pp. 267–275, 2005.

80. J. C. Santos, A. F. Portugal, F. D. Magalhães, and A. Mendes, "Optimization of medical PSA units for oxygen production," Industrial and Engineering Chemistry Research, vol. 45, no. 3, pp. 1085–1096, 2006.

81. K. G. Teague and T. F. Edgar, "Predictive dynamic model of a small pressure swing adsorption air separation unit," Industrial and Engineering Chemistry Research, vol. 38, no. 10, pp. 3761–3775, 1999.

82. K. T. Chue, J. N. Kim, Y. J. Yoo, S. H. Cho, and R. T. Yang, "Comparison of activated carbon and zeolite 13X for CO_2 recovery from flue gas by pressure swing adsorption," Industrial and Engineering Chemistry Research, vol. 34, no. 2, pp. 591–598, 1995.

83. J. H. Park, H. T. Beum, J. N. Kim, and S. H. Cho, "Numerical analysis on the power consumption of the PSA process for recovering CO_2 from flue gas," Industrial and Engineering Chemistry Research, vol. 41, no. 16, pp. 4122–4131, 2002.

84. A. L. Chaffee, G. P. Knowles, Z. Liang, J. Zhang, P. Xiao, and P. A. Webley, "CO_2 capture by adsorption: materials and process development," International Journal of Greenhouse Gas Control, vol. 1, no. 1, pp. 11–18, 2007.

85. J. S. D'amico, H. E. Reinhold III, and K. S. Knaebel, "Helium recovery," U.S. patent, 5, 542, 966, 1996.

86. N. K. Das, H. Chaudhuri, R. K. Bhandari, D. Ghose, P. Sen, and B. Sinha, "Purification of helium from natural gas by pressure swing adsorption," Current Science, vol. 95, no. 12, pp. 1684–1687, 2008.

87. A. P. G. Taveira and A. M. M. Mendes, "Xenon external recycling unit for recovery, purification and reuse of xenon in anesthesia circuits," U.S. patent 7, 442, 236, 2008.

88. S. Sircar and W. R. Kock, "Adsorptive separation of methane and carbon dioxide gas mixtures," European patent EP, 0193716, 1986.

89. S. Sircar, "High efficiency separation of methane and carbon dioxide mixtures by adsorption: adsorption and ion exchange," AIChE Symposium Series, vol. 84, pp. 70–72, 1988.

90. S. Sircar, R. Kumar, W. R. Koch, and J. Vansloun, "Recovery of methane from landfill gas," United States patent 4, 770, 676, 1988.
91. M. M. Davis, R. L. J. Gray, and K. Patei, "Process for the purification of natural gas," US patent, 5, 174, 796, 1992.
92. M. Mitariten, "Economic N_2 removal," Hydrocarbon Engineering, vol. 9, no. 7, pp. 53–57, 2004.
93. I. A. A. C. Esteves and J. P. B. Mota, "Simulation of a new hybrid membrane/pressure swing adsorption process for gas separation," Desalination, vol. 148, no. 17–3, pp. 275–280, 2002.
94. K. S. Knaebel and H. E. Reinhold, "Landfill gas: from rubbish to resource," Adsorption, vol. 9, no. 1, pp. 87–94, 2003.
95. C. A. Grande and A. E. Rodrigues, "Biogas to fuel by vacuum pressure swing adsorption I. Behavior of equilibrium and kinetic-based adsorbents," Industrial and Engineering Chemistry Research, vol. 46, no. 13, pp. 4595–4605, 2007.
96. C. A. Grande and R. Blom, "Utilization of Dual-PSA technology for natural gas upgrading and integrated CO_2 capture," Energy Procedia, vol. 26, pp. 2–14, 2012.
97. "Isosiv process operates commercially," Chemical & Engineering News, vol. 40, pp. 59–63, 1962.
98. T. C. Holcombe, "N-paraffin—isoparaffin separation process," U.S. patent, 4, 176, 053, 1979.
99. J. A. C. Silva, Separation of n/iso—paraffins by adsorption process [Ph.D. dissertation], University of Porto, Porto, Portugal, 1998.
100. A. Mersmann, B. Fill, R. Hartmann, and S. Maurer, "The potential of energy saving by gas phase adsorption processes," Chemical Engineering & Technology, vol. 23, pp. 937–944, 2000.
101. S. Sircar, "Pressure swing adsorption," Industrial and Engineering Chemistry Research, vol. 41, no. 6, pp. 1389–1392, 2002.
102. C. Voss, "Applications of pressure swing adsorption technology," Adsorption, vol. 11, no. 1, pp. 527–529, 2005.
103. K. Knaebel, "Adsorbent selection," 2004, http://www.adsorption.com/publications/ AdsorbentSel1B.pdf.
104. N. Sundaram and R. T. Yang, "On the pseudomultiplicity of pressure swing adsorption periodic states,"Industrial and Engineering Chemistry Research, vol. 37, no. 1, pp. 154–158,

1998.

105. W. A. Patrick, B. F. Lovelace, and E. B. Miller, "Method and apparatus for separating vapors and gases," U.S. patent 1, 335, 348, 1920.

106. A. B. Ray, "Process of recovering absorbable constituents from gas streams," U.S. patent 1, 548, 280, 1925.

107. R. T. Yang and P. L. Cen, "Improved pressure swing adsorption processes for gas separation: by heat exchange between adsorbers and by high-heat-capacity inert additives," Industrial & Engineering Chemistry Process Design and Development, vol. 25, no. 1, pp. 54–59, 1986.

108. H. Ahn, C. H. A. Lee, B. Seo, J. Yang, and K. Baek, "Backfill cycle of a layered bed H_2 PSA process," Adsorption, vol. 5, no. 4, pp. 419–433, 1999. · ·

109. W. D. Marsh, F. S. Pramuk, R. C. Hoke, and C. W. Skarstrom, "Pressure equalization depressurising in heatless adsorption," U.S. patent no. 3, 142, 547, 1964.

110. T. M. Stark, "Gas separation by adsorption process," U.S. patent 3, 252, 268, 1966.

111. M. P. S. Santos, C. A. Grande, and A. E. Rodrigues, "Pressure swing adsorption for biogas upgrading: effect of recycling streams in pressure swing adsorption design," Industrial and Engineering Chemistry Research, vol. 50, no. 2, pp. 974–985, 2011.

112. K. Warmuzinski, "Effect of pressure equalization on power requirements in PSA systems," Chemical Engineering Science, vol. 57, no. 8, pp. 1475–1478, 2002. · ·

113. J. A. Delgado and A. E. Rodrigues, "Analysis of the boundary conditions for the simulation of the pressure equalization step in PSA cycles," Chemical Engineering Science, vol. 63, no. 18, pp. 4452–4463, 2008.

114. J. L. Wagner, "Selective adsorption process," U.S. patent no. 3, 430, 418, 1969.

115. J. Xu, D. L. Rarig, T. A. Cook, K. K. Hsu, M. Schoonover, and R. Agrawal, "Pressure swing adsorption process with reduced pressure equalization time," US patent 6, 565, 628, 2003.

116. J. Xu and E. L. Weist Jr., "Six bed pressure swing adsorption process with four steps of pressure equalization," US patent 6,

454, 838, 2002.

117. N. Casas, J. Schell, and M. Mazzotti, "Pre-combustion CO_2 capture by PSA for IGCC plants," inProceedings of the 10th International Conference on Fundamentals of Adsorption (FOA '10), Awaji, Japan, May 2010.

118. F. V. S. Lopes, C. A. Grande, and A. E. Rodrigues, "Activated carbon for hydrogen purification by pressure swing adsorption: multicomponent breakthrough curves and PSA performance," Chemical Engineering Science, vol. 66, no. 3, pp. 303–317, 2011.

119. A. Fuderer, "Pressure swing adsorption with intermediate product recovery," US patent 4, 512, 780, 1985.

120. R. Ramachandran and L. H. Dao, "Process for recovering alkenes from cracked hydrocarbon streams," US patent 5, 744, 687, 1998.

121. F. A. Da Silva and A. E. Rodrigues, "Propylene/propane separation by vacuum swing adsorption using 13X zeolite," AIChE Journal, vol. 47, no. 2, pp. 341–357, 2001.

122. C. A. Grande and A. E. Rodrigues, "Propane/propylene separation by pressure swing adsorption using zeolite 4A," Industrial and Engineering Chemistry Research, vol. 44, no. 23, pp. 8815–8829, 2005.

123. A. D. Ebner, J. A. Ritter, M. D. LeVan, and J. C. Knox, "Unique regeneration steps for the sorbent-based atmosphere revitalization system designed for CO_2 and H_2O removal from spacecraft cabins," SAE International Journal of Aerospace, vol. 4, no. 1, pp. 488–493, 2011.

124. M. Yoshida, J. A. Ritter, A. Kodama, M. Goto, and T. Hirose, "Simulation of an enriching reflux PSA process with parallel equalization for concentrating a trace component in air," Industrial and Engineering Chemistry Research, vol. 45, no. 18, pp. 6243–6250, 2006.

125. K. P. Kostroski and P. C. Wankat, "High recovery cycles for gas separations by pressure-swing adsorption," Industrial and Engineering Chemistry Research, vol. 45, no. 24, pp. 8117–8133, 2006.

126. F. Dong, H. Lou, A. Kodama, M. Goto, and T. Hirose, "A new concept in the design of pressure-swing adsorption processes

for multicomponent gas mixtures," Industrial and Engineering Chemistry Research, vol. 38, no. 1, pp. 233–239, 1999.

127. F. Dong, H. Lou, A. Kodama, M. Goto, and T. Hirose, "The Petlyuk PSA process for the separation of ternary gas mixtures: exemplification by separating a mixture of CO_2-CH_4-N_2," Separation and Purification Technology, vol. 16, no. 2, pp. 159–166, 1999.

128. D. Basmadjian and A. L. Pogorski, "Process for the separation of gases by adsorption," US patent 3, 279, 153, 1966.

129. T. Tamura, "Absorption process for gas separation," US patent 3, 797, 201, 1974.

130. D. Diagne, M. Goto, and T. Hirose, "Parametric studies on CO_2 separation and recovery by a dual reflux PSA process consisting of both rectifying and stripping sections," Industrial and Engineering Chemistry Research, vol. 34, no. 9, pp. 3083–3089, 1995.

131. B. K. Na, H. Lee, K. K. Koo, and H. K. Song, "Effect of rinse and recycle methods on the pressure swing adsorption process to recover CO_2 from power plant flue gas using activated carbon," Industrial and Engineering Chemistry Research, vol. 41, no. 22, pp. 5498–5503, 2002.

132. S. P. Reynolds, A. D. Ebner, and J. A. Ritter, "Stripping PSA cycles for CO_2 recovery from flue gas at high temperature using a hydrotalcite-like adsorbent," Industrial and Engineering Chemistry Research, vol. 45, no. 12, pp. 4278–4294, 2006.

133. D. M. Ruthven, "PSA discussion," Studies in Surface Science and Catalysis, vol. 80, pp. 788–793, 1993.

134. R. Rota and P. C. Wankat, "Intensification of pressure swing adsorption processes," AIChE Journal, vol. 36, no. 9, pp. 1299–1312, 1990.

135. M. Whysall and L. J. M. Wagemans, "Very large-scale pressure swing adsorption processes," US patent 6, 210, 466, 2001.

136. W. D. Breck, Zeolite Molecular Sieves, John Wiley & Sons, New York, NY, USA, 1974.

137. J. E. Martin, M. T. Anderson, J. Odinek, and P. Newcomer, "Synthesis of periodic mesoporous silica thin films," Langmuir, vol. 13, no. 15, pp. 4133–4141, 1997.

138. D. Zhao, J. Feng, Q. Huo et al., "Triblock copolymer syntheses of mesoporous silica with periodic 50 to 300 angstrom pores,"

Science, vol. 279, no. 5350, pp. 548–552, 1998.
139. J. Rocha and M. W. Anderson, "Microporous titanosilicates and other novel mixed octahedral-tetrahedral framework oxides," European Journal of Inorganic Chemistry, no. 5, pp. 801–818, 2000.
140. S. Reyes, V. V. Krishnan, G. J. De Martin, J. H. Sinfelt, K. G. Strohmaier, and J. G. Santiesteban, "Separation of propylene from hydrocarbon mixtures," International patent, WO 03/080548 A1, 2003.
141. M. E. Rivera-Ramos, G. J. Ruiz-Mercado, and A. J. Hernández-Maldonado, "Separation of CO_2 from light gas mixtures using ion-exchanged silicoaluminophosphate nanoporous sorbents," Industrial and Engineering Chemistry Research, vol. 47, no. 15, pp. 5602–5610, 2008.
142. F. Rodríguez-Reinoso, M. Molina-Sabio, and M. T. González, "The use of steam and CO_2 as activating agents in the preparation of activated carbons," Carbon, vol. 33, no. 1, pp. 15–23, 1995.
143. Z. Liu, L. Ling, W. Qiao, and L. Liu, "Preparation of pitch-based spherical activated carbon with developed mesopore by the aid of ferrocene," Carbon, vol. 37, no. 4, pp. 663–667, 1999.
144. S. Jun, Sang Hoon Joo, R. Ryoo et al., "Synthesis of new, nanoporous carbon with hexagonally ordered mesostructure," Journal of the American Chemical Society, vol. 122, no. 43, pp. 10712–10713, 2000.
145. T. C. Golden, C. M. A. Golden, and D. P. Zwilling, "Self-supported structured adsorbent for gas separation," US patent 6, 565, 627, 2003.
146. J. L. C. Rowsell and O. M. Yaghi, "Metal-organic frameworks: a new class of porous materials," Microporous and Mesoporous Materials, vol. 73, no. 1-2, pp. 3–14, 2004.
147. P. D. C. Dietzel, Y. Morita, R. Blom, and H. Fjellvåg, "An in situ high-temperature single-crystal investigation of a dehydrated metal-organic framework compound and field-induced magnetization of one-dimensional metal-oxygen chains," Angewandte Chemie—International Edition, vol. 44, no. 39, pp. 6354–6358, 2005.

148. U. Mueller, M. Schubert, F. Teich, H. Puetter, K. Schierle-Arndt, and J. Pastré, "Metal-organic frameworks—prospective industrial applications," Journal of Materials Chemistry, vol. 16, no. 7, pp. 626–636, 2006. · ·

149. S. Ma, D. Sun, X. S. Wang, and H. C. Zhou, "A mesh-adjustable molecular sieve for general use in gas separation," Angewandte Chemie—International Edition, vol. 46, no. 14, pp. 2458–2462, 2007. · ·

150. S. Cavenati, C. A. Grande, A. E. Rodrigues, C. Kiener, and U. Müller, "Metal organic framework adsorbent for biogas upgrading," Industrial and Engineering Chemistry Research, vol. 47, no. 16, pp. 6333–6335, 2008. · ·

151. J. H. Cavka, S. Jakobsen, U. Olsbye et al., "A new zirconium inorganic building brick forming metal organic frameworks with exceptional stability," Journal of the American Chemical Society, vol. 130, no. 42, pp. 13850–13851, 2008. · ·

152. D. P. Valenzuela and A. L. Myers, Adsorption Equilibrium Data Handbook, Prentice Hall, New Jersey, NJ, USA, 1989.

153. R. T. Yang, Adsorbents. Fundamentals and Applications, John Wiley & Sons, New Jersey, NJ, USA, 2003.

154. G. Klein and T. Vermeulen, "Cyclic performance of layered beds for binary ion exchange," AIChE Symposium Series, vol. 71, no. 15, pp. 69–76, 1975.

155. M. Chlendi and D. Tondeur, "Dynamic behaviour of layered columns in pressure swing adsorption,"Gas Separation and Purification, vol. 9, no. 4, pp. 231–242, 1995. · ·

156. C. F. Watson, R. D. Whitley, and M. L. Meyer, "Multiple zeolite adsorbent layers in oxygen separation," US patent 5, 529, 610, 1996.

157. J. H. Park, J. N. Kim, S. H. Cho, J. D. Kim, and R. T. Yang, "Adsorber dynamics and optimal design of layered beds for multicomponent gas adsorption," Chemical Engineering Science, vol. 53, no. 23, pp. 3951–3963, 1998.

158. J. Yang and C. H. Lee, "Adsorption dynamics of a layered bed PSA for H_2 recovery from coke oven gas," AIChE Journal, vol. 44, no. 6, pp. 1325–1334, 1998.

159. Y. Lü, S. J. Doong, and M. Bülow, "Pressure-swing adsorption using layered adsorbent beds with different adsorption properties: I—results of process simulation," Adsorption, vol. 9, no. 4, pp. 337–347, 2003. · ·

160. S. Cavenati, C. A. Grande, and A. E. Rodrigues, "Separation of $CH_4/CO_2/N_2$ mixtures by layered pressure swing adsorption for upgrade of natural gas," Chemical Engineering Science, vol. 61, no. 12, pp. 3893–3906, 2006. · ·

161. C. A. Grande and A. E. Rodrigues, "Layered vacuum pressure-swing adsorption for biogas upgrading," Industrial and Engineering Chemistry Research, vol. 46, no. 23, pp. 7844–7848, 2007. · ·

162. C. A. Grande, S. Cavenati, and A. E. Rodrigues, "Separation column and pressure swing adsorption process for gas purification," World Patent Application, 2008/072215, 2008.

163. S. N. Vyas, S. R. Patwardhan, S. Vijayalakshmi, and K. S. Ganesh, "Adsorption of gases on carbon molecular sieves," Journal of Colloid And Interface Science, vol. 168, no. 2, pp. 275–280, 1994. · ·

164. R. Srinivasan, S. R. Auvil, and J. M. Schork, "Mass transfer in carbon molecular sieves-an interpretation of Langmuir kinetics," The Chemical Engineering Journal, vol. 57, no. 2, pp. 137–144, 1995.

165. S. Farooq, H. Qinglin, and I. A. Karimi, "Identification of transport mechanism in adsorbent micropores from column dynamics," Industrial and Engineering Chemistry Research, vol. 41, no. 5, pp. 1098–1106, 2002.

166. H. Qinglin, S. M. Sundaram, and S. Farooq, "Revisiting transport of gases in the micropores of carbon molecular sieves," Langmuir, vol. 19, no. 2, pp. 393–405, 2003. · ·

167. D. Shen, M. Bülow, and N. O. Lemcoff, "Mechanisms of molecular mobility of oxygen and nitrogen in carbon molecular sieves," Adsorption, vol. 9, no. 4, pp. 295–302, 2003. · ·

168. M. W. Ackley and R. T. Yang, "Diffusion in ion-exchanged clinoptilolites," AIChE Journal, vol. 37, no. 11, pp. 1645–1656, 1991.

169. L. Predescu, F. H. Tezel, and S. Chopra, "Adsorption of nitrogen, methane, carbon monoxide, and their binary mixtures on

aluminophosphate molecular sieves," Adsorption, vol. 3, no. 1, pp. 7–25, 1996.

170. W. Zhu, F. Kapteijn, J. A. Moulijn, M. C. Den Exter, and J. C. Jansen, "Shape selectivity in adsorption on the all-silica DD3R," Langmuir, vol. 16, no. 7, pp. 3322–3329, 2000. ··

171. D. H. Olson, "Light hydrocarbon separation using 8-member ring zeolites," US patent 6, 488, 741, 2002.

172. A. Jayaraman, A. J. Hernandez-Maldonado, R. T. Yang, D. Chinn, C. L. Munson, and D. H. Mohr, "Clinoptilolites for nitrogen/methane separation," Chemical Engineering Science, vol. 59, no. 12, pp. 2407–2417, 2004. ··

173. J. Gascón, W. Blom, A. van Miltenburg, A. Ferreira, R. Berger, and F. Kapteijn, "Accelerated synthesis of all-silica DD3R and its performance in the separation of propylene/propane mixtures," Microporous and Mesoporous Materials, vol. 115, no. 3, pp. 585–593, 2008. ··

174. S. M. Kuznicki, "Preparation of small-pored crystalline titanium molecular sieve zeolites," US patent 4, 938, 939, 1991.

175. S. M. Kuznicki, V. A. Bell, S. Nair et al., "A titanosilicate molecular sieve with adjustable pores for size-selective adsorption of molecules," Nature, vol. 412, no. 6848, pp. 720–724, 2001. ··

176. J. H. Wills, M. Shemaria, and M. J. Mitariten, "Production of pipeline-quality natural gas with the molecular gate CO_2 removal process," SPE Production and Facilities, vol. 19, no. 1, pp. 4–8, 2004.

177. R. P. Marathe, K. Mantri, M. P. Srinivasan, and S. Farooq, "Effect of ion exchange and dehydration temperature on the adsorption and diffusion of gases in ETS-4," Industrial and Engineering Chemistry Research, vol. 43, no. 17, pp. 5281–5290, 2004.

178. O. J. Smith IV and A. W. Westerberg, "Mixed-integer programming for pressure swing adsorption cycle scheduling," Chemical Engineering Science, vol. 45, no. 9, pp. 2833–2842, 1990.

179. S. Farooq, C. Thaeron, and D. M. Ruthven, "Numerical simulation of a parallel-passage piston-driven PSA unit," Separation and Purification Technology, vol. 13, no. 3, pp. 181–193, 1998. ··

180. F. A. Da Silva, J. A. Silva, and A. E. Rodrigues, "General package for the simulation of cyclic adsorption processes," Adsorption, vol. 5, no. 3, pp. 229–244, 1999. ··

181. L. T. Biegler, L. Jiang, and V. G. Fox, "Recent advances in simulation and optimal design of pressure swing adsorption systems," Separation and Purification Reviews, vol. 33, no. 1, pp. 1–39, 2004. · ·

182. P. A. Webley and J. He, "Fast solution-adaptive finite volume method for PSA/VSA cycle simulation; 1 single step simulation," Computers and Chemical Engineering, vol. 23, no. 11-12, pp. 1701–1712, 2000. · ·

183. L. Jiang, V. G. Fox, and L. T. Biegler, "Simulation and optimal design of multiple-bed pressure swing adsorption systems," AIChE Journal, vol. 50, no. 11, pp. 2904–2917, 2004. · ·

184. S. Nilchan and C. C. Pantelides, "On the optimisation of periodic adsorption processes," Adsorption, vol. 4, no. 2, pp. 113–147, 1998.

185. D. Nikolic, A. Giovanoglou, M. C. Georgiadis, and E. S. Kikkinides, "Generic modeling framework for gas separations using multibed pressure swing adsorption processes," Industrial and Engineering Chemistry Research, vol. 47, no. 9, pp. 3156–3169, 2008. · ·

186. D. Nikolic, E. S. Kikkinides, and M. C. Georgiadis, "Optimization of multibed pressure swing adsorption processes," Industrial and Engineering Chemistry Research, vol. 48, no. 11, pp. 5388–5398, 2009. · ·

187. V. Rama Rao, S. Farooq, and W. B. Krantz, "Design of a two-step pulsed pressure-swing adsorption-based oxygen concentrator," AIChE Journal, vol. 56, no. 2, pp. 354–370, 2010. · ·

188. N. Sundaram and P. C. Wankat, "Pressure drop effects in the pressurization and blowdown steps of pressure swing adsorption," Chemical Engineering Science, vol. 43, no. 1, pp. 123–129, 1988.

189. R. Kumar, "Adsorption column blowdown: adiabatic equilibrium model for bulk binary gas mixtures," Industrial and Engineering Chemistry Research, vol. 28, no. 11, pp. 1677–1683, 1989.

190. Z. P. Lu, J. M. Loureiro, A. E. Rodrigues, and M. D. LeVan, "Pressurization and blowdown of adsorption beds-II. Effect of the momentum and equilibrium relations on isothermal operation," Chemical Engineering Science, vol. 48, no. 9, pp. 1699–1707, 1993.

191. W. E. Waldron and S. Sircar, "Parametric study of a pressure swing adsorption process," Adsorption, vol. 6, no. 2, pp. 179–188, 2000. · ·
192. D. Ko, R. Siriwardane, and L. T. Biegler, "Optimization of a pressure-swing adsorption process using zeolite 13X for CO_2 sequestration," Industrial and Engineering Chemistry Research, vol. 42, no. 2, pp. 339–348, 2003.
193. A. Agarwal, L. T. Biegler, and S. E. Zitney, "A superstructure-based optimal synthesis of PSA cycles for post-combustion CO_2 capteffectively captureure," AIChE Journal, vol. 56, no. 7, pp. 1813–1828, 2010. · ·
194. K. Ramachandran, S. L. Lerner, and D. L. MacLean, "PSA multicomponent separation utilizing tank equalization," US patent 4, 816, 039, 1989.
195. A. Mehrotra, A. D. Ebner, and J. A. Ritter, "Arithmetic approach for complex PSA cycle scheduling,"Adsorption, vol. 16, no. 3, pp. 113–126, 2010. · ·
196. A. Mehrotra, A. D. Ebner, and J. A. Ritter, "Simplified graphical approach for complex PSA cycle scheduling," Adsorption, vol. 17, no. 2, pp. 337–345, 2011. ·
197. P. H. Turnock and R. H. Kadlec, "Separation of nitrogen and methane via periodic adsorption," AIChE Journal, vol. 17, pp. 335–342, 1971.
198. R. L. Jones, I. I. Keller, I. I. G. E, and R. C. Wells, "Rapid pressure swing adsorption process with high enrichment factor," US patent 4, 194, 892, 1980.
199. D. E. Earls and G. N. Long, "Multiple bed rapid pressure swing adsorption for oxygen," US patent 4, 194, 891, 1980.
200. T. J. Dangieri and R. T. Cassidy, "Enhanced performance in rapid pressure swing adsorption processing," W.O. patent 86/002015, 1986.
201. S. Sircar, "Gas separation by rapid pressure swing adsorption," US patent 5, 071, 449, 1991.
202. S. Sircar and B. F. Hanley, "Production of oxygen enriched air by rapid pressure swing adsorption,"Adsorption, vol. 1, no. 4, pp. 313–320, 1995. · ·

203. B. H. L. Betlem, R. W. M. Gotink, and H. Bosch, "Optimal operation of rapid pressure swing adsorption with slop recycling," Computers and Chemical Engineering, vol. 22, supplement 1, pp. S633–S636, 1998.

204. S. Kulish and R. P. Swank, "Rapid cycle pressure swing adsorption oxygen concentration method and apparatus," US patent 5, 827, 358, 1998.

205. B. G. Keefer, "High frequency pressure swing adsorption," U.S. Patent 6, 176, 897, 2001.

206. R. Arvind, S. Farooq, and D. M. Ruthven, "Analysis of a piston PSA process for air separation,"Chemical Engineering Science, vol. 57, no. 3, pp. 419–433, 2002. · ·

207. D. J. Connor, D. G. Doman, L. Jeziorowski et al., "Rotary pressure swing adsorption apparatus," US patent 6, 406, 523, 2002.

208. T. C. Golden, E. L. Weist Jr., and P. A. Novosat, "Adsorbents for rapid cycle pressure swing adsorption processes," US patent 7, 404, 846, 2008.

209. E. M. Kopaygorodsky, V. V. Guliants, and W. B. Krantz, "Predictive dynamic model of single-stage ultra-rapid pressure swing adsorption," AIChE Journal, vol. 50, no. 5, pp. 953–962, 2004. · ·

210. R. S. Todd and P. A. Webley, "Mass-transfer models for rapid pressure swing adsorption simulation,"AIChE Journal, vol. 52, no. 9, pp. 3126–3145, 2006. · ·

211. S. Alizadeh-Khiavi, J. A. Sawada, A. C. Gibbs, and J. Alvaji, "Rapid cycle syngas pressure swing adsorption system," US patent application 2007/0125228, 2007.

212. T. C. Golden and E. L. Weist, "Activated carbon as sole absorbent in rapid cycle hydrogen PSA," US patent 6, 660, 064, 2003.

213. M. J. LaBuda, T. C. Golden, R. D. Whitley, and C. E. Steigerwalt, "Performance stability in rapid cycle pressure swing adsorption systems," European Patent Application, vol. 1, pp. 917–994, 2008.

214. N. Sundaram, B. K. Kaul, E. W. Corcoran, C. Y. Sabottke, and R. L. Eckes, "Integration of rapid cycle pressure swing adsorption with refinery process units (hydroprocessing, hydrocracking, etc.)," US patent 7, 591, 879, 2009.

215. C. Siew-Wah, S. Sircar, and M. V. Kothare, "Miniature oxygen concentrators and methods," US patent 8, 226, 745, 2012.
216. F. V. S. Lopes, C. A. Grande, and A. E. Rodrigues, "Fast-cycling VPSA for hydrogen purification," Fuel, vol. 93, pp. 510–523, 2012.
217. S. Sircar, "Influence of gas-solid heat transfer on rapid PSA," Adsorption, vol. 11, no. 1, pp. 509–513, 2005. · ·
218. B. G. Keefer, A. Carel, B. Sellars, I. Shaw, and B. Larisch, "Adsorbent laminate structures," US patent 6, 692, 626, 2004.
219. B. G. Keefer and C. R. McLean, "High frequency rotary pressure swing adsorption: Apparatus," US patent 6, 056, 804, 2000.
220. A. S. T. Chiang and M. C. Hong, "Radial flow rapid pressure swing adsorption," Adsorption, vol. 1, no. 2, pp. 153–164, 1995. · ·
221. J. Smolarek, F. W. Leavitt, J. J. Nowobilski, V. E. Bergsten, and J. H. Fassbaugh, "Radial bed vaccum/pressure swing adsorber vessel," US patent 5, 759, 242, 1998.
222. W. C. Huang and C. T. Chou, "Comparison of radial- and axial-flow rapid pressure swing adsorption processes," Industrial and Engineering Chemistry Research, vol. 42, no. 9, pp. 1998–2006, 2003.
223. D. M. Ruthven, "Past progress and future challenges in adsorption research," Industrial and Engineering Chemistry Research, vol. 39, no. 7, pp. 2127–2131, 2000.
224. P. Xiao, J. Zhang, P. Webley, G. Li, R. Singh, and R. Todd, "Capture of CO_2 from flue gas streams with zeolite 13X by vacuum-pressure swing adsorption," Adsorption, vol. 14, no. 4-5, pp. 575–582, 2008. · ·
225. S. Dasgupta, N. Biswas, Aarti et al., "CO_2 recovery from mixtures with nitrogen in a vacuum swing adsorber using metal organic framework adsorbent: A Comparative Study," The International Journal of Greenhouse Gas Control, vol. 7, pp. 225–229, 2012.
226. Z. Liu, C. A. Grande, P. Li, J. Yu, and A. E. Rodrigues, "Multi-bed vacuum pressure swing adsorption for carbon dioxide capture from flue gases," Separation and Purification Technology, vol. 81, pp. 307–317, 2011.
227. C. Shen, Z. Liu, P. Li, and J. Yu, "Two-stage VPSA process for CO_2 capture from flue gas using activated carbon," Industrial & Engineering Chemistry Research, vol. 51, pp. 5011–5021, 2012.

228. C. A. Grande, F. Poplow, and A. E. Rodrigues, "Vacuum pressure swing adsorption to produce polymer-grade propylene," Separation Science and Technology, vol. 45, no. 9, pp. 1252–1259, 2010. · ·

229. M. Tagliabue, D. Farrusseng, S. Valencia et al., "Natural gas treating by selective adsorption: material science and chemical engineering interplay," Chemical Engineering Journal, vol. 155, no. 3, pp. 553–566, 2009. · ·

230. C. A. Grande and R. Blom, "Utilization of dual-PSA technology for natural gas upgrading and integrated CO_2 capture," Energy Procedia, vol. 26, pp. 2–14, 2012.

231. S. Sircar, "Separation of multicomponent gas mixtures," US patent, 4, 171, 206, 1978.

232. Y. Chen, A. Kapoor, and R. Ramachandran, "Two stage pressure swing adsorption process," US patent 5, 993, 517, 1999.

233. Y. Takamura, S. Narita, J. Aoki, S. Hironaka, and S. Uchida, "Evaluation of dual-bed pressure swing adsorption for CO_2 recovery from boiler exhaust gas," Separation and Purification Technology, vol. 24, no. 3, pp. 519–528, 2001. · ·

234. D. L. Rarig, T. C. Golden, and E. L. Weist Jr., "Purification of CO_2 from H_2 PSA vent gas," in Proceedings of the National AIChE Meeting, Indianapolis, Ind, USA, November 2002.

235. C. A. Grande and R. Blom, "Dual pressure swing adsorption units for gas separation and purification," Industrial & Engineering Chemistry Research, vol. 51, pp. 8695–8699, 2012.

Chapter 7

The Role of Natural Fractures in Shale Gas Production

Ian Walton[1] and John McLennan[1]

[1]Energy and Geoscience Institute, University of Utah, USA

ABSTRACT

Natural fractures seem to be ubiquitous in shale gas plays. It is often said that their presence is one of the most critical factors in defining an economic or prospective shale gas play. Many investigators have presumed that open natural fractures are critical to gas production from deeper plays such as the Barnett, as they are for shallower gas shales such as the Devonian shales of the northeastern US and for coal bed methane plays. A common view on production mechanisms in shales is "because the formations are so tight gas can be produced only when extensive networks of natural fractures exist" [6]. However, there is now a growing body of evidence that any natural fractures that do exist may well be filled with calcite or other minerals and it has even been suggested that open natural fractures would in fact be detrimental to Barnett shale gas production [9].

Commercial exploitation of low mobility gas reservoirs has been improved with multi-stage hydraulic fracturing of long horizontal wells. Favorable results have been associated with large fracture surface area in contact with the shale matrix and it is here that the role of natural fractures is assumed to be critical. For largely economic reasons hydraulic fracturing for increasing production from shale gas reservoirs is often carried out using large volumes of slickwater injected at pressures/rates high enough to create and propagate extensive hydraulic fracture systems. The fracture systems are often complex, due essentially to intersection of the hydraulic fractures with the natural fracture network. After hydraulic fracturing operations the injected water is flowed back. Typically, only a small percentage (on the order of 20 to 40%) is recovered.

In this paper we investigate the role played by natural fractures in the gas production process. By applying a new model of the production process to data from many shale gas wells across a number of shale plays in North America, we can for the first time begin to sort out assertion from inference in the role that these fractures play. Specifically, we are able to estimate the magnitude of the fracture surface through which gas is actually produced. We are able to demonstrate that although it may be commensurate with the expected surface area of open natural fractures for the ultra-low permeability shallow gas shales, it is in fact commensurate with a very much smaller area for the deeper gas shales such as the Barnett. Furthermore, given a typical value of the matrix permeability, almost all the gas between the fractures would have been produced in an uncharacteristically short period of time unless the producing fractures are 100s of feet apart. The implications of these findings for completion and stimulation strategies will be discussed.

INTRODUCTION

One of the main premises of our investigation of the production processes in shale gas plays is that the industry's mental picture of the process remains very much influenced by the concepts developed in the 1990s of the production processes in coal bed methane resources and in shallow shale gas plays. It is appropriate therefore that we begin our discussion of shale gas production characteristics by reviewing this early work.

Coal Bed Methane

Coal is a heterogeneous and anisotropic porous medium, characterized by two distinct porosity systems:

- Micropores of diameter of the order of nanometers with almost zero permeability.
- Macropores or cleats, slot-like with spacing of the order of 2 cm and width of the order of microns; permeability is stress-dependent but far in excess of the micro-pore permeability. They are often formed by shrinkage of the coal matrix due to dewatering during the coalification process.

Gas is stored essentially by adsorption in the coal matrix; very little is stored as free gas in the micropores or as free gas or dissolved gas in the connate brine in the cleats. In the subsurface the cleats are usually filled with water, some of which must be produced to the surface to facilitate gas production.

The conventional view of the production process divides it into three stages (see Figure 1):

- Cleat dewatering, lasting of the order of several years; the water production rate gradually falls as water is removed from the cleats. At the same time more and more gas is produced at increasing rates and the relative permeability to gas in the cleats increases leading to lower pressures and more gas production.
- Stabilized flow: eventually, most of the water in the cleats has been removed, the cleat fluid pressure bottoms out and the relative permeability to gas levels off. Over this period the gas rate slowly peaks.

- Decline: there is then no more increase in drawdown available to sustain gas production and gas production declines. If the cleat pressure was constant and the pseudo-steady-state (PSS) regime is applicable (see the later discussion for definitions of this flow regime), then gas production rate should decline exponentially.

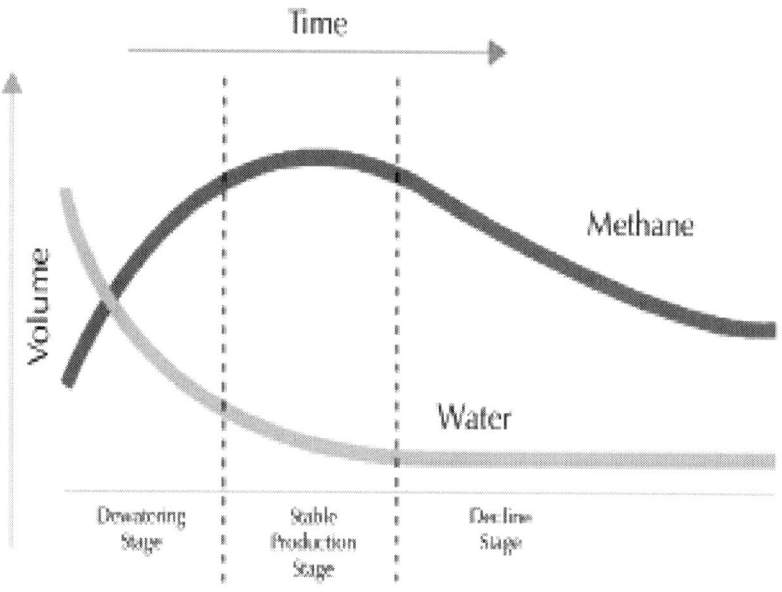

Figure 1: Stages in gas and water production from coal (after [1]).

There are three essential elements to a model of the CBM production process:

- Transport in the coal matrix, modeled as a diffusive process using Fick's diffusion law. In principle the gas concentration in the coal matrix satisfies a diffusion equation, but it is common to use a pseudo-steady-state (PSS) approximation similar to that proposed by Warren and Root [2] in their dual porosity formulation of production from naturally fractured reservoirs. For example, King et al [3] used the PSS simplification to reduce computing time and because after a period of time the numerical accuracy was deemed to be quite acceptable. We have estimated from King's data that the PSS solution is valid beyond about 40 days, which is much shorter than the typical duration of the production process. We note, however, that this time scale depends on the assumed values of the diffusivity in the matrix and on the spacing of the cleats, assumed to be of the order of a few cm.

- Desorption at the cleat/matrix interface as characterized by the Langmuir isotherm
- Transport of water and free gas in the cleat system. To avoid difficulties in defining the configuration of the cleat system, it is common to adopt a dual porosity description in which the cleat system is treated as a continuum with system characteristics analogous to those of a porous medium. Two-phase flow in this system can for the most part be adequately described by Darcy flow. In narrower cleats it may be necessary to include capillary pressure and slippage effects especially at low pressure.

It is apparent that the cleat or natural fracture system plays a very important part in the production process. The density of the cleats plays two critical roles: first, the close spacing of the cleats reduces the time required for the gas to diffuse to the cleats and, second, it is associated with a high cleat/matrix surface area without which economical gas production would be unlikely. The width of the cleat is a primary influence on the pressure drop in the whole system and therefore on the water and gas production rates. In situ the cleats are usually water-filled and presumably kept open by the pressure of the fluid they contain. The cleats may close somewhat as the pressure falls during production, though this may be more than offset by matrix shrinkage as the gas is desorbed.

Devonian Shales of the Appalachian Basin

Gas production from Devonian shales received a great deal of attention in the 1980s and early 1990s as a result of US DoE initiatives. This is well documented in many GRI reports and industry publications. The consensus view is that these reservoirs are highly fractured containing a substantial number of fractures with spacing of the order of 1-10 cm (see, for example, [4]). Luffel et al [5] measured the matrix permeability at less than 0.1 nd. Water content of the Devonian Shale averages 2.5 to 3% of bulk volume and appears to be at irreducible water saturation. Typical depth is a few thousand feet, pore pressure is less than about

3000 psi and about 50% of the gas in place is adsorbed; there is little or no water production.

Carlson and Mercer [6] summarized the consensus view of the production process as "because the formations are so tight gas can be produced only when extensive networks of natural fractures exist." The extent to which this statement holds for other gas shale plays is debatable, but it has certainly been influential in developing the industry's vision of what is happening downhole.

Gatens et al [7] used a dual-porosity model similar to that formulated by Warren and Root [2] but extended to use the unsteady-state equation instead of the pseudo-steady-state (PSS) equation for matrix flow. Analysis of hundreds of Devonian wells showed that most of the production data fell into the linear transient regime (as we discuss later in this document). Luffel et al [5] obtained a good history match with data by assuming an open fracture spacing of a few feet, while Carlson and Mercer [6] needed a fracture spacing of about 80 ft, both with matrix permeability of less than 0.1 nd. An issue that does not seem to have been addressed however, is whether these fractures, if present, are in fact open and if so how they are maintained open against closure stress. If, like coal cleats, they are initially water-filled, is water production observed? An implicit assumption seems to be that they are open and gas-filled.

Carlson and Mercer [6] proposed that molecular diffusion is the dominant transport mechanism in the matrix in these extremely tight reservoirs, in which case a matrix diffusion coefficient should be used instead of the matrix permeability. They did not evaluate the consequences of this hypothesis. It remains a possibility that the use of such a coefficient would reduce the need for a large fracture surface area and ultimately for the need to propose the existence of a large open fracture network.

Thus there are three essential elements to a model of the production process in the Devonian shales:

- Desorption of gas in the matrix (as characterized typically by the Langmuir isotherm)
- Transport in the matrix towards the fracture network, modeled as Darcy flow even though the permeability is extremely small.
- Transport of free gas in the fracture system.

Devonian Antrim Shale of the Michigan Basin

The Antrim Shale is a shallow, under-pressured, naturally-fractured shale reservoir with characteristically low matrix permeability, and with adsorbed gas, free gas and mobile water co-existing in the reservoir. A typical Antrim well will produce considerable quantities of water early in life, and as dewatering of the reservoir progresses, water production rates decline and a corresponding increase in gas production is normally observed (as a result of gas desorption with reduced reservoir pressures), similar to a CBM well. In fact, the Antrim shale is often considered to be a hybrid of productive dry gas shale and CBM plays. It has characteristics which are similar to these other unconventional reservoirs, but it is also different in many ways. The Antrim shale is more intensely fractured than the Devonian Shales of the Appalachian Basin, with fracture spacing as close as 1 to 2 ft. Kuuskraa et al [8] have noted that the "intensity and interconnection of the fractures govern the shale's natural producibility."

The typical depth of the Antrim shale is less than about 2000 feet, pore pressure is a few hundred psi and more than 70% of the gas in place is adsorbed, the remainder being stored essentially as free gas in the matrix pores. Peak gas may occur as late as 3 years into production. Production data has been history-matched using similar software to that used for CBM [8]. It was found that fracture spacing of the order of a few inches facilitated a good match with production data. It was stated that if a fracture spacing of 3-6 ft was used (which is compatible with observations from cores and logs), then production would be an order of magnitude lower than observed in existing wells. One possible resolution of this conundrum may lie, as the authors suggest, in detrital silt layers within the matrix that could provide conductive flow paths. (An alternative explanation that remained unconsidered by the authors lies in the use of the PSS approximation for matrix transport, which may be completely invalid in this context.)

This description leaves many issues unresolved, but importantly places the Antrim shale as an intermediate between CBM and the other shales, in that the natural fractures appear to be conductive and initially water-filled, but has free pore gas in addition to adsorbed gas.

Deeper Gas Shales

Most modern commercial gas shale plays are similar in many respects to the Barnett shale, though there are of course many differences and variations in the values of the parameters that control the gas production process. In these relatively deep and high pressure reservoirs, most of the gas is stored as pore gas, but the production process is still similar to that described above for shallower shale gas plays.

The subject of open natural fractures is one of the most contentious within the community of Barnett workers. Many investigators have presumed that open natural fractures are critical to Barnett gas production, as they are for the shallower gas shales, even though there is now a growing body of evidence that any natural fractures that do exist may well be filled with calcite or other minerals (see Figure 2). There are also arguments that suggest that if there was an abundance of open natural fractures within the Barnett, there would be a much smaller gas accumulation present within the reservoir. Open natural fractures, if they existed, would have led to major expulsion and migration of gas out of the shale into overlying rocks, substantially decreasing pore pressure within the Barnett and, hence, the amount of gas in place. The Barnett would not be over-pressured (that is, over-pressured relative to the bounding strata) if copious open natural fractures existed. Note that the Barnett is not just the gas reservoir, but also the source, trap, and seal for the gas; if the seal is fractured and inefficient, then the present gas in place would be reduced because the free gas would be lost, and only the adsorbed gas would remain in the shale (a similar situation to that of the Antrim Shale of northern Michigan). The huge amount of gas in place, in an over-pressured and fully-saturated (in terms of sorption) state, is ultimately what makes the Barnett so prolific.

Figure 2: Mineralized natural fractures in a Barnett shale sample (adapted from [9]).

A common argument for the necessity for open natural fractures in shale gas plays is that a large surface area is necessary for economic gas production from these very tight rocks. Later in this paper we analyze production data to estimate the magnitude of the fracture surface through which gas is actually produced. We are able to demonstrate that although it may be commensurate with the expected surface area of open natural fractures for the ultra-low permeability shallow gas shales, it is in fact commensurate with a very much smaller area for the deeper gas shales such as the Barnett. Furthermore, given the typical permeability of the Barnett shale (some 100 times that of the shallow gas shales), almost all the gas between the fractures would have been produced in an uncharacteristically short period of time unless the fractures are 100s of feet apart. These issues and conclusions will be discussed at length later in the paper.

PRODUCTION MECHANISMS, PRODUCTION MODELING TECHNIQUES AND SIMULATORS

Having outlined the pertinent characteristics of unconventional gas reservoirs, we now document the likely production mechanisms in the various shale plays based on our understanding of their geology and the underlying geophysics.

The matrix permeability of shale gas reservoirs is extremely small, probably on the order of one tenth of a microdarcy or 100nd. It is virtually impossible to produce gas from these reservoirs in commercial quantities unless the wells are hydraulically fractured and even then, or so it is commonly believed, production is really only possible because a network of natural fractures is opened up. (It is interesting to note that gas has been produced from the ultra-tight Devonian shale plays of the North Eastern USA from more conventionally-fractured vertical wells, which implies that multi-stage hydraulic fracturing was unnecessary for these plays. This is the first hint that the role of the natural fractures may be quite different for the Devonian plays and the deeper shale plays.)

An essential element of a mathematical model of gas production from shales is therefore the ability to describe flow in a very tight rock matrix and flow in a network of fractures. In most gas shale reservoirs most of the reservoir fluid is stored in the matrix and the primary flow path is from the matrix into the fractures and thence into the wellbore. There are essentially two methods of characterizing a multiply-fractured reservoir:

- Discrete fracture network (DFN) model, in which the fractures are defined explicitly in terms of their location in the reservoir, their connectivity to one another and to the wellbore and their production characteristics, such as permeability and conductivity.
- Dual porosity/dual permeability models in which the fracture network is treated as a continuum in much the same way as a porous medium is treated as a continuum for analysis of flow characteristics.

Discrete Fracture Network Models

Many commercial numerical reservoir simulators have the capability of simulating flow through a complex network consisting of pores and fractures. However, one of the greatest drawbacks and limitations of simulating a discrete fracture network model is the a priori assumption that all relevant properties of the fracture network are known. Nevertheless great insights can be obtained into the impact of the essential physical processes by examining simple fracture configurations. We note that in principle many different physical and

petro-physical components can be included in numerical simulations. However, in practice it is quite common to see results presented only for the special cases:
- Reservoir fluid of small and constant compressibility.
- Production under constant drawdown conditions.
- No desorption.
- Darcy flow in fractures and matrix.
- Matrix and fracture permeability independent of pressure; it is often assumed that fracture conductivity is essentially infinite.

The simplest fracture network that has been applied to shale gas production consists of a number of planar fractures placed transversely to a horizontal wellbore as illustrated schematically in Figure 3. It is apparent from many published numerical studies that under these circumstances flow from the reservoir can be described in terms of a number of identifiable flow regimes. The following account is taken from a recent paper by Luo et al [10]. These authors used a commercial reservoir simulator to calculate the flow into a horizontal well with six infinitely-conductive transverse fractures as shown in Figure 4.

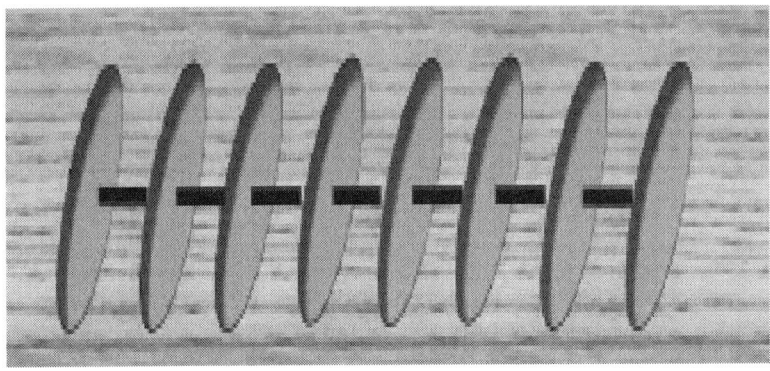

Figure 3: Idealized discrete fracture network showing multiple transverse fractures originating from a horizontal well.

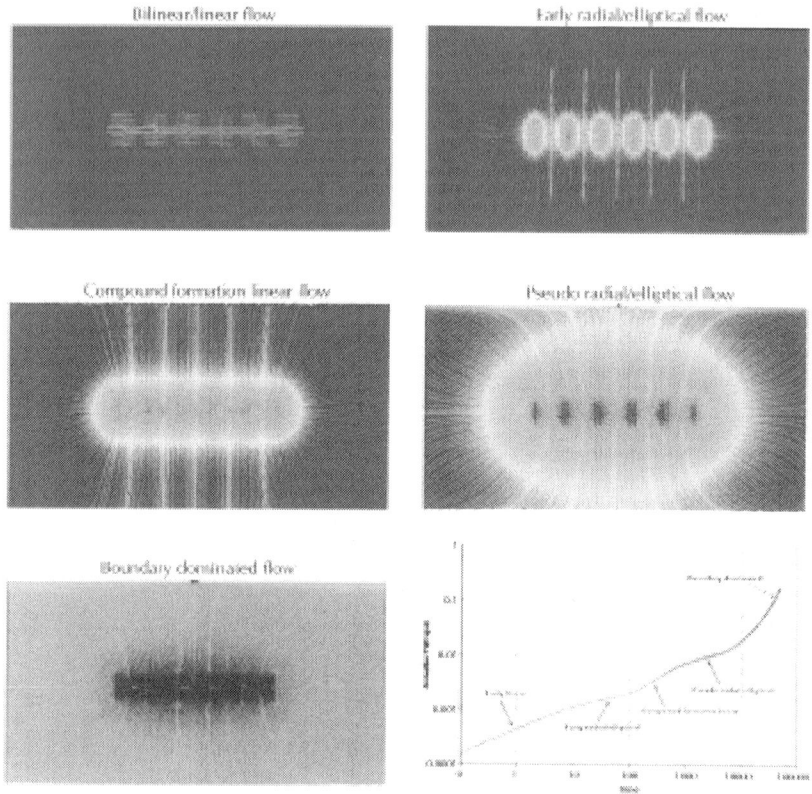

Figure 4: Numerical solutions for flow into six infinitely-conductive transverse fractures (taken [10]).

The flow behavior can be conveniently discussed in terms of five flow regimes as follows.

- Bilinear or linear flow: soon after the well is placed on production reservoir fluid flows normal to the fracture planes and along the fractures into the well. The streamlines are shown in yellow in Figure 4; reservoir pressure is in red and the constant bottomhole or well pressure is in blue. Note that flow into the fracture tips is negligible and each fracture behaves independently of the other fractures. This regime may also be termed the infinite-acting regime in the sense that the neighboring fractures are effectively at infinity. The duration of this regime depends, as we shall see later, on many parameters including the matrix permeability, the fluid compressibility and the fracture spacing. The illustrations in Figure 4 are for infinite conductivity fractures.

- Early radial/elliptical flow: flow into the fracture tips is present, but weak; flow into the fracture surfaces is still predominantly linear, but fracture interference is just beginning to impact the flow. Note that the pressure drawdown in the matrix has almost reached the mid-line between the fractures. At this point the flow regime may be described as pseudo-steady-state or fracture-boundary-dominated.
- Compound formation linear flow (CFL flow): here the fractures are fully interactive and the reservoir drainage area is dominated by the area defined by the length of the well and the length of the fractures. Flow from beyond this area grows in importance.
- Pseudo-radial/elliptical flow: flow from beyond the wellbore/fracture area grows in importance and appears to be radial or elliptical.
- Reservoir-boundary-dominated flow: ultimately the outer boundary of the reservoir begins to impact the flow.

It is difficult to infer from these simulations the time scales and duration of these flow regimes for parameter values other than those used in the particular simulation. Indeed, this highlights one of the severe drawbacks to the full numerical approach to modeling production flow in these reservoirs: it is difficult to make general conclusions about the characteristics of the flow and their dependence on the input data without undertaking very many numerical simulations; this is a formidable task even for a restricted input data space. However, based on the semi-analytic models that are described below, we believe that for many shale gas wells it would be unusual to expect to encounter the compound formation linear flow regime for at least 10 years after the well was placed on production.

Dual Porosity/Dual Permeability Models

The conventional view of a naturally-fractured reservoir is that it is a complex system composed of irregular matrix blocks surrounded by a network of more highly permeable fractures. In reality in tight gas shales some or most of the fractures may not be open to flow or they are opened up only during the hydraulic fracturing process. Warren and Root [2] were among the first to recognize that the simple model of reservoir flow based on single values of the permeability and porosity does not apply to naturally-fractured reservoirs, though they had in mind reservoirs quite different from gas shale reservoirs. In order to handle the problems associated with lack of detailed information on the structure of the fracture network they proposed a dual-porosity model in which a primary porosity associated with inter-granular pore spaces is augmented by a secondary porosity related to that of the network of natural fractures. At each point in space there are two overlapping continua—one for the matrix and one for the fracture network. The detailed geometry of the fracture system need not be specified in this model, but can include as particular examples any of the discrete fracture models described above. In typical shale gas applications the matrix has high storage capacity but low permeability and the fractures have relatively low storage capacity and higher permeability. It is quite possible (or, indeed, likely) that in many shale gas reservoirs no gas is stored in the fractures, though they may become filled with frac fluid during the hydraulic fracturing process.

In the dual porosity formulation flow from the matrix to the fractures is described by a transfer function with Darcy flow characteristics. The original Warren-Root models incorporated the pseudo-steady-state assumption in the matrix blocks and assigned a single value to the pressure within the blocks; the mass transfer rate from the matrix to the fractures depends then on the pressure differential between the matrix and the fracture. Thus these models assumed, almost implicitly, that sufficient time had elapsed that the flow in the matrix blocks between the fractures was already fracture-boundary-dominated. Later in this paper we estimate the time scale on which inter-fracture pseudo-steady-state begins and find that it is typically of the order of several years for a fracture spacing of 100 ft or more. This is quite consistent with typical simulation results described above. If the fracture spacing was as small as 10ft, we should expect to see fracture interference or the onset of PSS flow after about 10 days. For shale gas reservoirs, more complex models (unsteady state or fully transient) are needed to resolve the flow in the matrix in more detail.

A detailed discussion of the Warren-Root model, its background and similar contemporary models can be found in the excellent monograph by van Golf-Racht [11]. We note in particular that Kazemi [12] was one of the first to extend the Warren-Root model to include transient flow in the matrix blocks. Some seventeen years after Warren and Root published their seminal work, Kucuk and Sawyer [13] adapted their model for flow in shale reservoirs by incorporating effects such as desorption from the organic matrix material and Knudsen flow in the pores and, of course, incorporating full transient effects in the matrix blocks.

In the years following the formulation of the dual porosity model for naturally fractured reservoirs, solutions of the coupled partial differential equations for the pore and fracture fluid pressures were obtained using finite difference techniques. While these simulations can provide accurate solutions, often in a complicated geometry covering the entire reservoir, the large number of computations involved made them cumbersome for analysis of large data sets. In response, an alternative, faster, method of solution was developed in the 1980s. For a simplified geometry, Laplace transform solutions were developed, in which the transformed solutions were inverted numerically, using, typically, the Stehfast algorithm.

Several authors have noted that analytic approximations can be developed for certain ranges of parameter values (referring to the Warren-Root dimensionless parameters defined below) appropriate for shale gas reservoirs. It will become apparent later in this paper that for typical shale gas reservoirs the interporosity flow coefficient (or transmissivity), λ, is very small and this allows asymptotic approximations to be derived for the Laplace-transformed solutions. Since these models still require numerical inversion of the transformed solution, it would be more accurate to label them semi-analytic models. They have advantages over full numerical solutions in terms of speed of calculation and in the added value they bring to understanding the flow characteristics and the impact of the reservoir and completion parameters on production.

Development of New Semi-analytic Solution

Recently, we have taken the idea of developing asymptotic solutions one stage further. We have developed perturbation solutions for $\lambda<<1$ directly from the dual porosity partial differential equations, thereby removing the necessity for Laplace transforms altogether. The greater simplicity and enhanced understanding afforded by these solutions will become apparent as we proceed. (Full details are available in an internal EGI report [14].) The result is similar to the simple linear flow model that is currently gaining favor in the literature, but has some notable advantages:

- The model does not make the *a priori* assumption of linear flow into a sequence of transverse fractures.
- The model does not make the *a priori* assumption of infinite fracture conductivity, and allows an estimate to be made for the fracture pressure loss.
- Identification of the end of the linear infinite-acting flow period and development of the ensuing PSS solution is made explicit.
- Provision of a solution form that facilitates fast and easy production data analysis.
- Identification of the reservoir and completion parameters that are the greatest (primary) determinants of productivity.
- Solution scheme that permits rational extension to include other physical processes, such as desorption.

For simplicity we restrict attention in this paper to single-phase flow in the matrix and assume that gas is produced at constant bottom hole pressure; we shall also neglect the impact of gas desorption. For the purposes of the present discussion the most important part of the solution is the leading order solution for the reservoir pseudo-pressure, which satisfies a standard diffusion equation.

$$\frac{\partial^2 m_{Dm_0}}{\partial z_D^2} = \frac{\partial m_{Dm_0}}{\partial t_D} \tag{1}$$

Here we have used dimensionless parameters (denoted by subscript D); full definitions are provided in the Nomenclature section later in this paper. Dimensionless distance normal to the fracture face, z_D, is scaled on L/2 (i.e. half the inter-fracture spacing), and dimensionless time, t_D, is scaled on a time scale, that characterizes pressure diffusion in the matrix

$$t_m = \frac{c\varphi_m \mu (L/2)^2}{k_m} \quad (2)$$

The appropriate initial and boundary conditions are

$$m_{Dm_0} = 0 \quad at \quad t_D = 0$$

$$m_{Dm_0} = m_{Df_0} = 1 \quad at \quad z_D = 0, \quad \frac{\partial m_{Dm_0}}{\partial z_D} = 0 \quad at \quad z_D = 1 \quad (3)$$

The leading order influx from the matrix into the fracture network is given by

$$q_{D0} = \left| \frac{\partial m_{Dm_0}}{\partial z_D} \right|_{z_D = 0} \quad (4)$$

At downhole conditions the mass flowrate from the matrix into the fracture network is

$$q_m = q_{ch} q_{D0} \quad (5)$$

The dimensionless flowrate is defined in equation (4) and the characteristic mass flowrate is defined by

$$q_{ch} = A \frac{M k_m}{2 R T_w} \frac{m_{ch}}{Z_{ch}} \quad (6)$$

Here A denotes the productive fracture surface area. The total mass flow rate measured at surface conditions is

$$q_s = \frac{q_m}{\rho_s} = A \frac{k_m}{Z_{ch}} \frac{T_s}{T_w} \frac{Z_s}{P_s} m_{ch} \qquad (7)$$

Here subscript s denotes surface conditions. Note that q_s is actually a volumetric flowrate and is measured typically in units such as scf/s.

The dimensionless flowrate defined in equation (4) may be readily calculated in terms of the dimensionless pseudo-pressure, either from the full numerical solution of the diffusion equation, (equation (1)), or from the early-time infinite-acting approximation to it. Both solutions provide very useful information and insights into the variation of the production rate with time. The dimensionless cumulative production is defined by

$$Q_{D0} = \int_0^{t_D} q_{D0} \, dt_D \qquad (8)$$

The diffusion equation (1) is readily solved using standard numerical schemes as made available in mathematical software such as MATLAB. To complement this solution we have obtained an analytic approximation valid while the change in pressure has not been impacted by neighboring boundaries or fractures—often referred to as the infinite-acting approximation. The early-time approximation to the dimensionless inflow rate is

$$q_{D0} = \frac{1}{\sqrt{\pi t_D}} \qquad (9)$$

And the early-time approximation to the dimensionless cumulative production is

$$Q_{D0} = 2\sqrt{\frac{t_D}{\pi}} \qquad (10)$$

Figure 5 compares the full and early-time solutions for the dimensionless flow rate and cumulative production against dimensionless time. For convenience we have used a log-log plot here. There are several major features of this plot that are worthy of further comment:

- The early-time (infinite-acting) solution provides a good approximation to the full numerical solution for dimensionless time, t_D, less than about 0.15, which translates to about 3 years in dimensional terms.
- During this time frame, the flowrate is represented by a straight line of slope – ½ and the cumulative production is represented by a straight line of slope + ½.
- Boundaries (or in this case neighboring fractures) begin to influence the flow after this point in time and the solution departs from the simple linear dependence on the square-root of time. This is also evident in the numerical solutions presented by Bello and Wattenbarger [15] in theirFigures 5, 7, 8 and 10 and in the field data shown in their Figure 1.
- The dimensionless cumulative production approaches a final value of 1 as it should because of the way we have defined the characteristic scales in our non-dimensionalization of the equations.
- The error incurred in estimating the cumulative production by extrapolating the infinite-acting solution beyond its region of validity is apparent from Figure 5.
- Almost 90% of the total gas that can be produced has been produced by the time $t_D=1$. This then provides a simple interpretation of the matrix diffusion time as the time (in real terms) to produce 90% of the gas available.

Figure 6 shows the full and infinite-acting solutions for cumulative production plotted against the square-root of dimensionless time. As expected from equation (10) the early-time solution is well represented by a straight line with slope $2/\sqrt{\pi}$ or 1.128. Later in this paper we develop this plot as the basis of our production data interpretation technique.

In anticipation of the application of these results to analysis of production data, it is useful to provide an expression for the cumulative production in dimensional terms. Analogous to equation (6) we define cumulative production at downhole conditions by

$$Q_m = Q_{ch} Q_{D0} \qquad (11)$$

The characteristic cumulative production scale is defined as

$$Q_{ch} = q_{ch} t_m = A \frac{Mk_m}{2RT_w} \frac{m_{ch}}{z_{ch}} t_m \qquad (12)$$

The time scale t_m is defined in equation (2) and Q_{D0} is defined in equation (8).

Analogous to equation (9) we define cumulative production at surface conditions by

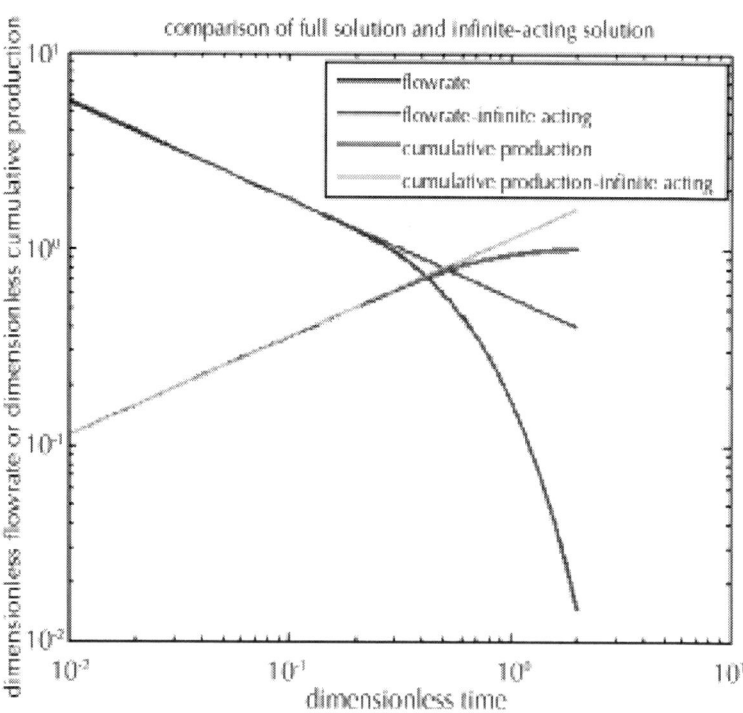

Figure 5: A comparison of the full and early-time solutions for the dimensionless flow rate and cumulative production.

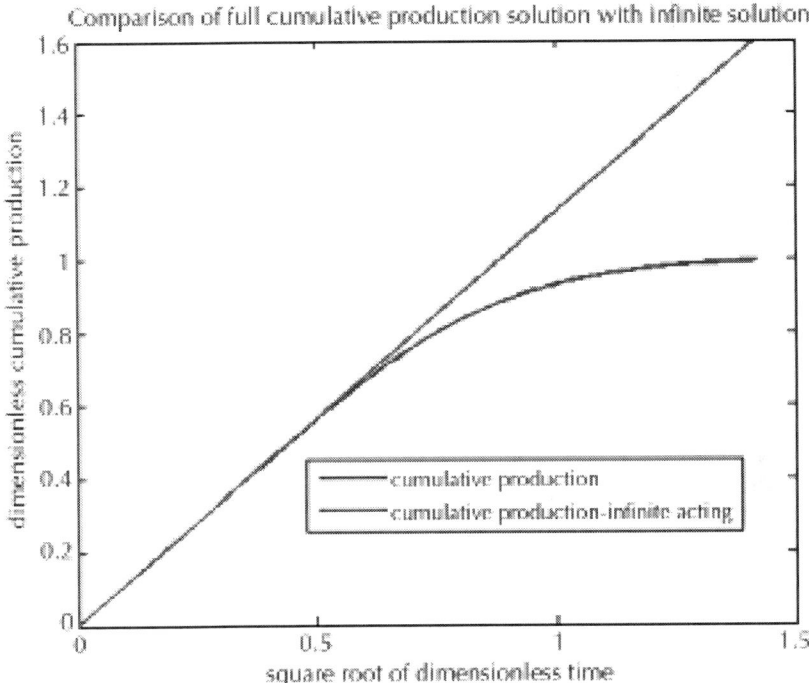

Figure 6: A comparison of the full and early- time solutions for the dimensionless cumulative production.

In view of the wide applicability of the early-time solution, it is useful to state the form taken by equation (13) during the infinite-acting period. Using the early-time approximation given in equation (9), we see that

$$Q_s = A \frac{k_m}{Z_{ch}} \frac{T_s}{T_w} \frac{Z_s}{p_s} m_{ch} \, 2\sqrt{\frac{t_m}{\pi}} \sqrt{t} \qquad (14)$$

If we now use the definition of T_m (equation (2)), we can express the cumulative production at surface conditions as

$$Q_s = C_p \sqrt{t} \qquad (15)$$

Where

$$Q_s = A \frac{k_m}{Z_{ch}} \frac{T_s}{T_w} \frac{Z_s}{p_s} m_{ch} 2 \sqrt{\frac{t_m}{\pi}} \sqrt{t}$$

(16)

The coefficient C_p represents the slope of the dimensional equivalent of the straight line in Figure 6 and is of fundamental importance in the subsequent development of this paper. For now we will observe that C_p characterizes the early-time solution in a form that is easy to estimate from production data. We shall refer to it as the "Production Coefficient". (We have adopted this terminology in recognition of the similarity of this result to a well-known expression for the leakoff rate of a compressible fluid from a fracture to a reservoir filled with the same fluid (see, for example, [15]).

PRODUCTION DATA INTERPRETATION

The analysis described above suggests that for a substantial part of a shale gas well's production history, the cumulative production varies linearly with the square-root of time. The coefficient C_p represents the slope of the straight line in a plot of Q against √t and characterizes the early-time solution in a form that is easy to compare with production data The time scale, T_m, defined inequation (2) defines the upper limit of the applicability of the linear flow regime and allows us to characterize the production rate once boundary-effects have become important.

We illustrate this production analysis technique by comparing production from a group of wells in the Barnett shale. Figure 7 shows a conventional plot of production rate against time for several wells that had been producing for at least 5 years in an area of the Barnett field. The data was obtained from a public data base and we have plotted the production rates at yearly intervals. For clarity of presentation we have connected the data points by smooth lines. This "conventional" plot reveals nothing about the relative decline rates of the wells or provides much insight into the flow regime(s). The same data sets have been plotted in the new format in Figure 8. It is immediately apparent that for most of these wells the data falls on straight lines as expected from our analysis. The slope of these lines is readily measured and provides an estimate for the production coefficient, C_p. Estimation of C_p is quick and easy and provides us with a new metric with which we can quantify the productivity of these wells. Again, we explore these results in more detail later, but for now we note that the linear flow regime extends beyond at least 5 years, since there is no indication at this point of departure from the straight line in this plot. This fact alone sets some bounds on the fracture spacing and the matrix permeability.

In Figure 8 we have omitted the first year's cumulative production data. Generally, early-time production data is quite severely impacted by variable drawdown conditions and so we should not expect a good straight line fit at that time. Analysis of this regime is discussed at length elsewhere [14].

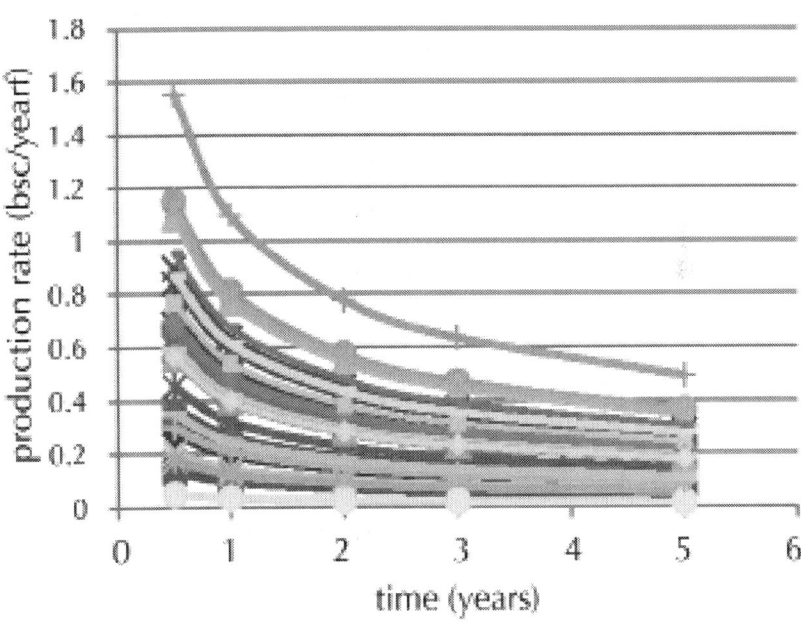

Figure 7: Production data from a group of Barnett wells plotted in the conventional format.

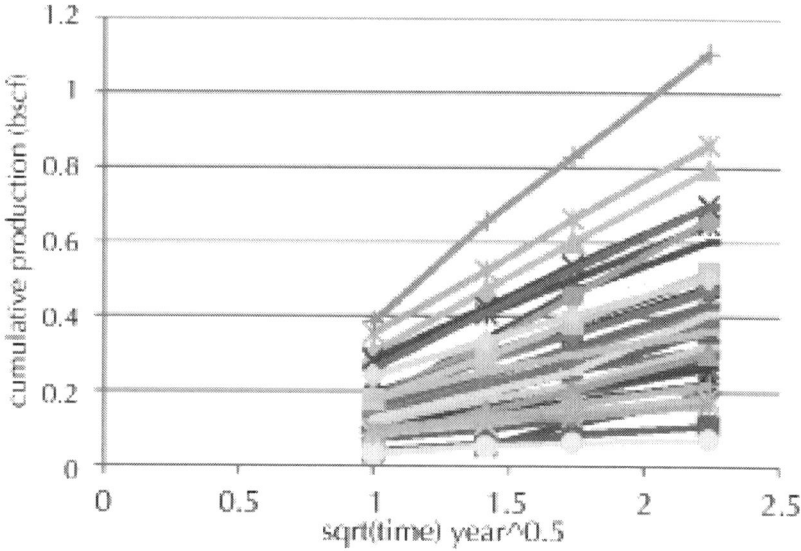

Figure 8: Same production data from Figure 7 plotted in the new format.

IDENTIFICATION OF PRODUCTION DRIVERS: NATURE VERSUS NURTURE

We have established the applicability of this new analysis technique to data from very many wells in many shale gas plays across the US and Canada, though there is neither space nor time available to discuss this in detail in the present paper. Following on from that analysis, we will now go on to discuss the results in more depth and begin to draw some tentative conclusions about the production drivers, by examining the parameters that together constitute the formula for C_p in equation (15). We may divide these parameters broadly into two groups—those that characterize the nature of the reservoir and those that characterize how we "nurture" the reservoir. Specifically the parameters are:

Nature:
- Matrix permeability, k_m
- Matrix porosity, φ_m

- Gas viscosity, μ
- Gas compressibility, c
- Initial reservoir pressure, p_i
- Reservoir temperature, T

Nurture:
- Bottom hole flowing pressure, p_w
- Productive fracture surface area, A

In developing these results we are constrained by the requirement that λ<<1, where

$$\lambda = \frac{12 k_m r_w^2}{L\, c_f} \tag{17}$$

This requirement sets some bounds on the fracture network characteristics, but these are generally easily met for shale gas reservoirs. For given values of the matrix permeability and the wellbore radius, the combination of fracture spacing and fracture conductivity must be sufficiently large.

It is apparent that given these conditions production for a large part of the production history of these wells depends upon the parameters listed above. We note in particular that history matching production data over this flow period furnishes only one parameter and that is the production coefficient, C_p. That is all. The square-root of time behavior is inherent to the physics of the flow: i.e. linear flow into a network of (effectively infinitely-conductive) fractures. It is not at all surprising that conventional history-matching techniques using reservoir simulators give non-unique answers: many different values of the parameters in the list above can together constitute the same value of the production coefficient. Moreover this formulation tells us what parameters have little effect on the history-matching process, including the precise value of the fracture conductivity. Even if history-matching were attempted in terms of dimensionless parameters, it is apparent that the result is insensitive to λ provided that it is small enough.

We need to elaborate at this point on the parameter A defined above as the "productive fracture surface area". This is the area of the fractures in contact with the reservoir that serve as the channels that convey gas from the matrix to the wellbore. We can make no assertion at this point about whether these fractures are natural fractures or propped or unpropped hydraulic fractures and nor can we say anything (yet) about their spacing or their lengths or indeed their number and location. *All we can infer from the production data analysis is the total productive fracture surface area.*

Some insight about the spacing of these productive fractures can be obtained by examining the time scale of pressure diffusion in the matrix. We demonstrated earlier that we may expect the root-time solution to be valid until neighboring fractures begin to compete with one another for production. In other words, until pressure diffusion in the matrix can no longer be considered to be independent of the fracture spacing. According to the analysis presented above we should expect the cumulative data to deviate from a straight-line in the root-time plot for $t > 0.15\, t_m$.

If we could detect the time at which this departure occurs then, we have some information with which to estimate the productive fracture spacing. Even if the entire production history to date is in the linear flow regime, we can make an estimate of a lower bound on the fracture spacing. As we see later, the fracture spacing is surprisingly large for typical shale gas plays.

We have now analyzed many shale gas production data sets using our proposed technique and have found the square-root fit to be very good. Based on this and on the mathematical analysis that supports that technique we have concluded that the production rate declines inversely with the square root of time. As we have discussed above, this is a consequence of the dominant production process of linear flow into a network of fractures. The decline rate is therefore fully determined by the physics of the production process. We should not expect to see any significant variation from well to well, from vertical to horizontal wells or indeed form play to play. What does vary is the multiplier, the production coefficient, C_p, which as we have demonstrated elsewhere depends on many factors, principally the reservoir quality, the reservoir and bottom hole pressure and the productive fracture surface area.

Example: Barnett Shale Production Data Analysis

As we have indicated above, it is relatively straightforward to use this new technique to analyze production data whether it is on a well-by-well basis or averaged over a play or area within a play. In essence, the process consists of three steps:

- From the daily (or monthly or yearly) production data calculate the cumulative production for each well at different points in time.
- Plot cumulative production against the square-root of time.
- Estimate the Production Coefficient form the slope of the best straight line fit to the data.

The Barnett shale is a good starting point for a more in-depth data analysis, since production data is readily available from public databases and, moreover, that data extend over many wells for long periods of time. The Barnett shale occupies several counties in North Texas. It is broadly bounded by geologic and structural features and may be divided according to estimates of maturity into a gas window and oil window. Historically the major development has been in a core area located to the north of Fort Worth (Figure 9), but more recently expansion has occurred to the south and to the west. To date many thousands of wells have been drilled and completed in the Barnett, initially vertical, but now almost entirely horizontal.

It has become common practice to sub-divide the Barnett play into three areas, described as the Core, Tier 1 and Tier 2. For convenience we may define these areas according to county as follows

- Core region: Denton, Tarrant and Wise counties, comprising 2974 horizontal wells and 3886 vertical wells
- Tier 1 region: Hood, Johnson and Parker counties, comprising 3865 horizontal wells and 251 vertical wells
- Tier 2 region: all other counties, comprising 687 horizontal wells and 401 vertical wells

It is apparent form this cursory division that the fraction of wells that were completed horizontally shifts from 43% in the Core area to 94% in the Tier 1 area, which reflects the development of technology with time and the spread of drilling with time to the outer areas. At the date of these figures (2009) Tier 2 was relatively unexploited.

The result of this detailed analysis (Figure 10) allows us to quantify the production variations in the Barnett Core, Tier 1 and Tier 2 areas and to distinguish the impact of horizontal and vertical well completions on the productivity. In a sense this represents a first, somewhat crude, pass at distinguishing the impact of nature (in the sense that reservoir properties depend on location, with the core area providing more fertile ground than Tier 1 or Tier 2) and nurture (in the anticipation that horizontal well technology provides more productive fracture surface area than does vertical well technology).

Figure 9: Development of the Barnett shale in North Texas.

In Figure 10 we have shown the cumulative distribution of production coefficient for each of the six categories defined above. The plots should be interpreted as follows. For each category the probability that a well has a specified value of the production coefficient in excess of the value on the x-axis can be read off the y-axis. For example the probability of a horizontal well in the Core having a production coefficient in excess of 0.75 (bcf/yr^0.5) is about 8%.

It is apparent that, as is to be expected, wells in the Core have better production characteristics than wells in Tier 1 and wells in Tier 2 and that in general horizontal wells have better production characteristics then vertical wells. It is interesting in this context to examine the variation of production coefficient in the core area in more detail. Figure 11 shows the location of ten of the wells in the core area with high values of the production coefficient (in green), 10 of the wells with medium values (in blue) and 10 wells with low values (in red). It is apparent that there appear to be sweet spots even within the core area, but there are substantial outliers and there are some relatively poor wells close to better wells.

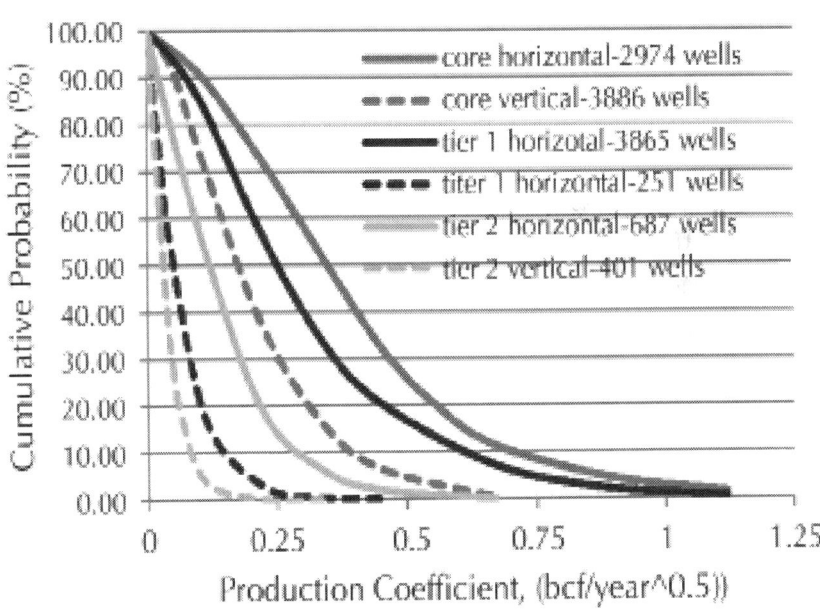

Figure 10: Distribution of values of the Production Coefficient for horizontal and vertical wells in the Core region, Tier 1 region and Tier 2 region of the Barnett shale.

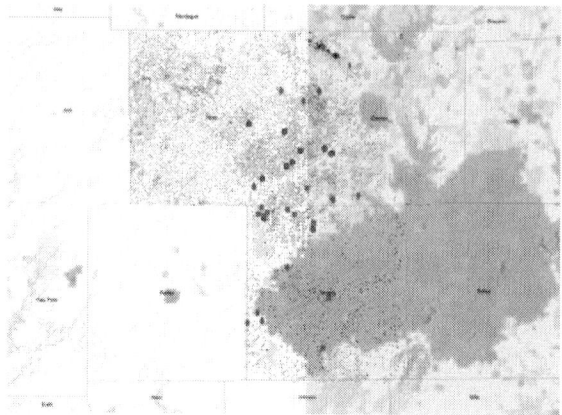

Figure 11: Preliminary identification of sweet spots in the core area of the Barnett. Each black dot represents a well; 10 wells with high Production Coefficients are identified in green, 10 medium performers are in blue and 10 poor performers are identified in red.

IMPLICATIONS AND DEDUCTIONS

One of the advantages of the semi-analytic method outlined in this paper is that it enables us to make certain deductions about the magnitude of the parameters that drive the productivity of the well. In particular we can make some inferences about the magnitude of the fracture surface area through which the gas is produced and about the likely spacing of the productive fractures.

Productive Fracture Surface Area

Our analytic solution allows us to relate the Productivity Coefficient C_p to a group of parameters that may be roughly divided into those that characterize the nature of the reservoir and those that characterize the impact of the completion and stimulation strategy. In our formula for C_p (equation (16)) perhaps the parameter that has the greatest uncertainty is the productive fracture surface area, A. This is the area of the fractures in contact with the reservoir that serve as the channels that convey gas from the matrix to the wellbore. We can make no assertion at this time point about whether these fractures are natural fractures or propped or unpropped hydraulic fractures and nor can we say anything (yet) about their spacing or their lengths or indeed their number and location. We can, however, make an estimate from the production data analysis of the total surface area of these productive fractures.

In Figure 12 we show estimates of the productive fracture surface area for typical strong, medium and weak performing wells in the core area of the Barnett in terms of values of the matrix permeability. In this calculation we have made reasonable estimate of the other parameters that impact the productivity coefficient such as the gas viscosity and compressibility and the matrix porosity.

Several features of this plot merit discussion:
- Well productivity increases with productive fracture surface area and with matrix permeability, as expected.
- Wells in this group typically have matrix permeabilities in the range 100-300 nd and this implies that the productive fracture surface area is in the range 1-6 million square feet (Msqft): the higher the permeability, the less fracture surface area is needed to achieve the given productivity.
- If the matrix permeability was as low as even 10 nd, the required fracture surface area would approach 100 Msqft. On the other hand, matrix permeabilities of the order of 1 µd would require less than about 1 Msqft of productive fracture surface area.

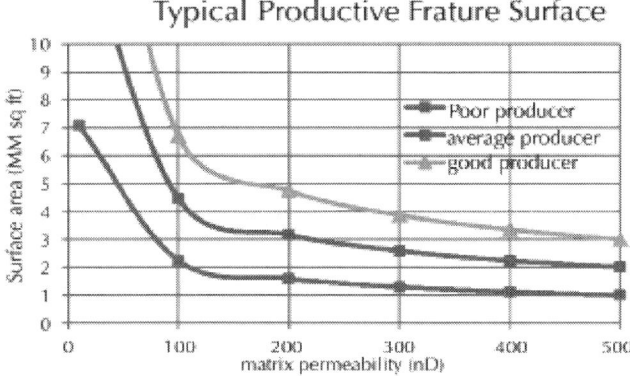

Figure 12: Estimate of productive fracture surface area for specified matrix permeability. The three curves were developed based on analysis of Barnett shale well data using the EGI semi-analytic production model.

To place these numbers in context we note that a fracture of height 200 ft and half-length 200 ft has surface area of 0.16 Msqft. Thus, 20 of these fractures would have a fracture surface area of 3.2 Msqft, which is a perfectly plausible estimate of the hydraulic fracture surface area created with modern multi-stage fracturing techniques. (Note that 20 such fractures would be spaced about 150 ft apart in a 3000 ft lateral.)

Figure 12 was developed on the basis of production data from wells in the core area of the Barnett shale, but the results apply, at least qualitatively, to other shale or tight gas plays. For example for more conventional tight gas plays for which the permeability is of the order of 1 μd or more, we should expect respectable productivity with only one such hydraulic fracture, which reinforces our experience that a vertical well with a single bi-wing fracture may be adequate for those reservoirs, but not for shale gas plays. Conversely the productive fracture surface area for economic production from ultra-tight shale plays (such as the shallow shale plays described earlier in this paper) cannot be achieved by producing from the hydraulic fractures alone.

We have been careful so far to make no formal distinction between the natural fractures and the hydraulic fractures in so far as productivity is concerned. All we have demanded is that their conductivity is sufficiently large that $\lambda<<1$, which should not in principle present too great a restriction on the fracture conductivity whether it is associated with propped fractures or unpropped fractures. On the basis of our data analysis we should expect a productive fracture surface area in the range of 1-6 Msqft. How then is that area created and what are the implications of this figure?

A crude estimate of the fracture surface area that is created by pumping large volumes of frac fluid may be made by performing a mass balance and assuming that none of the fluid has leaked off or imbibed into the formation over the time in which the fracture network is created. For a total fluid volume V and assuming a created average fracture width w during pumping, the total fracture surface area may be estimated at (using any consistent set of units of course).

$$A = 2\frac{V}{w}$$

If, for example, 100 million gallons of frac fluid were pumped and the assumed frac width was 0.2 inches, then the total surface area would be about 100 Msq ft. Clearly, this is far in excess of our estimate of the productive fracture surface area and would suggest that less than 10% of the created fracture surface area is actually productive. Naturally, this raises all sorts of other questions concerning the efficiency of this process, which we plan to address in a future project.

A similar mass balance for the proppant placed in a typical job enables an estimate to be made of the surface area of propped fractures. If we make some estimate of the likely width (0.1 in) and porosity (0.4) of a propped fracture (after closure), it appears that a propped fracture surface area of the order of a few million square feet is quite plausible.

Productive Fracture Spacing

Some insight about the spacing of these productive fractures can be obtained by examining the time scale of pressure diffusion in the matrix. We demonstrated earlier that we may expect the root-time solution to be valid until neighboring fractures begin to compete with one another for production. In other words until, pressure diffusion into a fracture can no longer be considered to be independent of the fracture spacing. According to the analysis presented earlier we should expect the cumulative production data to deviate from a straight-line in the root-time plot for $t>0.15T_m$. If we could detect the time at which this departure occurs then, we have some information with which to estimate the productive fracture spacing. Even if the entire production history to date is in the linear flow regime, we can at least make an estimate of a lower bound on the fracture spacing.

It is instructive to estimate the matrix diffusion time for typical values of the fracture spacing and matrix permeability. The results are shown in Figure 13. The diffusion time increases quadratically with the fracture spacing and inversely with the matrix permeability. Typical values for the diffusion time are quite low. For example, if, as we expect, linear flow continues for at least 3 years, then we should expect to see a diffusion time of the order of 20 years. Figure 13 suggests that the productive fracture spacing is likely to be of the order of 100 ft or more.

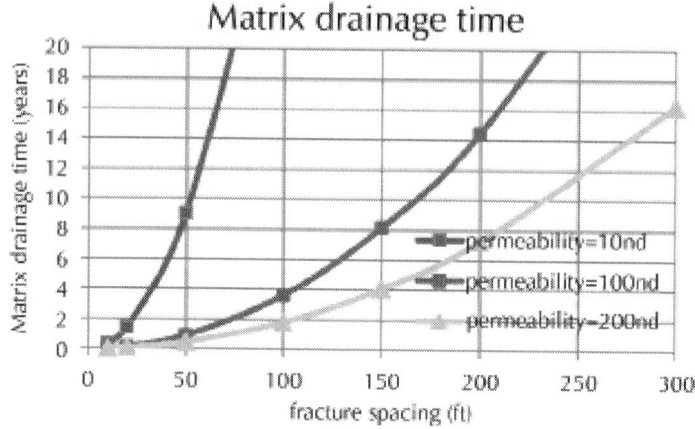

Figure 13: The impact of fracture spacing on the time to produce 90% of the gas in place.

In figure 13 we have identified the diffusive time scale with the matrix drainage time. As we showed above, 90% of the total gas in the pore space between the fractures has been drained by this time. A time scale of about 20 years is at least consistent with the industry estimates of the effective production lifetime of these wells. It is worth noting here the consequences of much smaller fracture spacing. For a fracture spacing of only 10 ft, we estimate that 90% of the total gas production will have occurred within the first few months of production, which is quite unrealistic. Note also that the surface area of planar fractures only 10 ft apart in a 3000 ft lateral would be of the order of 150 Msq ft, which again is unreasonably large.

CONCLUSIONS

A common view of production mechanisms in shales is "because the formations are so tight gas can be produced only when extensive networks of natural fractures exist" [6]. To this extent gas production from some of the shallower (Devonian) shales is similar to gas production from coal. As we have discussed earlier in this paper, we expect that the deeper gas shales differ in this respect.

Using a new semi-analytic production model, we have analyzed production data from a number of shale gas wells in several different

North American shale gas plays. Interpretation of the results suggest that productivity is largely determined by a small group of parameters that may be decomposed into two sub-groups representing the nature of the reservoir (such as matrix permeability and porosity) and what we may term our (engineering) attempts at nurture (including completion and stimulation parameters). Of key importance is the productive fracture surface, which unfortunately is difficult to estimate a priori. However, our interpretation of the production data suggest the following

- Productive fracture surface area ~1-6 Msqft and probably within 2-4 Msq ft.
- The volume of these productive fractures is very much less than the volume of water pumped, but
- Productive fracture volume scales approximately with the volume of proppant placed.
- Typically, there is no indication of fracture interference during production even after several years, which suggests that the productive fracture spacing is at least 100 ft.
- Time to drain 90% of the fractured region or matrix blocks: ~10-20 years

We are led to the conclusion that almost all the fracturing fluid pumped during a multi-stage horizontal well fracturing operation in the shales serves to open a vast, and possibly complex, network of natural fractures and that these fractures do not make a significant contribution to the well's productivity. We are led inevitably to questions concerning the conductivity of these, largely unpropped, fractures and to investigate the rock and fluid mechanisms that seemingly prevent them from being productive. The role of the fracturing fluid (usually slickwater) in this process should now be investigated from this new perspectivel

REFERENCES

1. A Guide to Coalbed Methane Reservoir Engineering, Saulsberry, J.L., Schafer, P.S., and Schraufnagel, R.A. (Editors), Gas Research Institute Report GRI-94/0397, Chicago, Illinois (March 1996).
2. Warren, J.E. and Root, P.J.: "The Behavior of Naturally Fractured reservoirs," SPEJ, September 1963. (Originally published as SPE 00426, 1962).

3. King, G.R, Ertekin, T, and Schwerer, F.C., "Numerical simulation of the Transient Behavior of Coal- Seam Degasification Wells," SPE Formation Evaluation, April, 1986.
4. Schettler, P.D., Parmely, C.R. and Lee, W.J., "Gas Storage and Transport in Devonian Shales." SPEFE, September 1989.
5. Luffel, D.L., Hopkins, C.W. and Schettler, P.D., "Matrix Permeability Measurement of Gas Productive Systems,", SPE 26633 (1993).
6. Carlson, E.S. and Mercer, J.C., "Devonian Shale Gas production: mechanisms and Simple Models," SPE 19311, 1989 (also JPT April 1991).
7. Gatens, J.M., Lee, W.J., and Rahim, Z.: "Application of an Analytic Model to History Match Devonian Shales Production Data," Paper SPE 14509 presented at the 1985 Eastern Regional meeting, Morgantown, W Virginia, November 6-8, 1985.
8. Kuuskraa, V.A., Wicks, D.E. and Thurber, J.L.: "Geologic and Reservoir Mechanisms Controlling Gas Recovery from the Antrim Shale," Paper SPE 24883 presented at the 67th Annual SPE Technical Conference and Exhibition, Washington, D.C., October 4-7, 1992.
9. Kent Bowker, HAPL Technical Workshop, 2008)
10. Luo, S., Neal, L., Arulampalam, P. and Ciosek, J.M.: "Flow Regime Analysis of Multi-stage Hydraulically-fractured Horizontal Wells with Reciprocal Rate Derivative Function: Bakken case Study," Paper CSUG/SPE 137514 presented at the Canadian Unconventional Resources and International Petroleum Conference, Calgary, Alberta, Canada, 19-21 October, 2010.
11. Van Golf-Racht, T.D.: "Fundamentals of Fractured Reservoir Engineering," Developments in Petroleum Science, vol 10, Elsevier Scientific Publishing Company, 1982.
12. Kazemi, H.: "Pressure Transient Analysis of naturally Fractured Reservoirs with Uniform Fracture Distribution," SPEJ (Dec 1969), 451-61; Trans AIME, 246
13. Kucuk, F. and Sawyer, W.K.: "Transient Flow in Naturally Fractured Reservoirs and Its Application to Devonian Gas Shales," Paper SPE 9397 presented at the 55th Annual Technical Conference and Exhibition, Dallas, Texas, September 21-24 1980.

14. Walton, I.C.: "Shale Gas Production Analysis, Phase I Final Report," EGI internal report 100983, 2012.
15. Bello, R.O. and Wattenbarger, R.A.: "Rate Transient Analysis in Naturally Fractured Shale gas Reservoirs," Paper SPE 114591 presented at the CIPC/SPE Gas Technology Symposium, Calgary, Alberta, June 16-19, 2008.
16. Economides, M.J. and Nolte, K.G.:"Reservoir Stimulation," Prentice Hall, Third Edition, 2000.

Citations

CHAPTER 1

Annick Nago and Antonio Nieto, "Natural Gas Production from Methane Hydrate Deposits Using Clathrate Sequestration: State-of-the-Art Review and New Technical Approaches," Journal of Geological Research, vol. 2011, Article ID 239397, 6 pages, 2011. doi:10.1155/2011/239397.

CHAPTER 2

Mohamed Iqbal Pallipurath, "Effect of Bed Deformation on Natural Gas Production from Hydrates," Journal of Petroleum Engineering, vol. 2013, Article ID 942597, 9 pages, 2013. doi:10.1155/2013/942597.

CHAPTER 3

Rolando Barrera, Carlos Salazar, and Juan F. Pérez, "Thermochemical Equilibrium Model of Synthetic Natural Gas Production from Coal Gasification Using Aspen Plus," International Journal of Chemical Engineering, vol. 2014, Article ID 192057, 18 pages, 2014, doi:10.1155/2014/192057.

CHAPTER 4

Kenneth Kekpugile Dagde and Jackson Gunorubon Akpa, "Numerical Simulation of an Industrial Absorber for Dehydration of Natural Gas Using Triethylene Glycol," Journal of Engineering, vol. 2014, Article ID 693902, 8 pages, 2014. doi:10.1155/2014/693902.

CHAPTER 5

Qin He, Shahab D. Mohaghegh, and Vida Gholami, "A Field Study on Simulation of CO_2 Injection and ECBM Production and Prediction of CO_2 Storage Capacity in Unmineable Coal Seam," Journal of Petroleum Engineering, vol. 2013, Article ID 803706, 8 pages, 2013. doi:10.1155/2013/803706.

CHAPTER 6

Carlos A. Grande, "Advances in Pressure Swing Adsorption for Gas Separation," ISRN Chemical Engineering, vol. 2012, Article ID 982934, 13 pages, 2012. doi:10.5402/2012/982934.

CHAPTER 7

Ian Walton and John McLennan (2013). The Role of Natural Fractures in Shale Gas Production, Effective and Sustainable Hydraulic Fracturing, Dr. Rob Jeffrey (Ed.), ISBN: 978-953-51-1137-5, InTech, DOI: 10.5772/56404.

Index

A

Acid gases recovery unit (AGR) 46, 48
Axisymmetric framework 15

C

Carbon conversion efficiency (CCE) 72
Carbon molecular sieves (CMS) 146
Coal bed methane production (CBM/ECBM) 115
Cold gas efficiency (CGE) 44
Computational fluids dynamic model (CFD) 42
Computer Modeling Group Ltd. (CMG) 132
Coupled pore fluid 23

D

Diffusion 119
Discrete fracture network (DFN) 180

E

Energy demand 16
Enhanced coal bed methane (ECBM) 115
Equivalence ratio (ER) 54, 61

F

Fossil fuel 116
Future work 82

G

Gas hydrates 3, 4, 6

Global efficiencie 40, 76, 80, 82
Global population 2

H

High quality 71, 82

I

Integrated gasification combined cycle plant (IGCC) 42

M

Methyl diethanolamine (MDEA) 91
Mobility gas 172
Monoethanolamine (MEA) 91
Morphology 17
Multicomponent mixture of gase 145

N

Natural gas 90
Natural gas hydrate (NGH) 16
Natural gas hydrates 3

P

Partial pressure 146
Pressure swing adsorption (PSA) 135, 137, 143
Production Coefficient 192, 197, 199
Pseudo-pressure 186, 188
Pseudo-steady-state (PSS) 174, 176

Q

Quite unrealistic 204

R

Relative error (RE) 66
Root mean square deviation (RMSD) 66
Root mean square error (RMSE) 66

S

Simulation of cumulative gas 25
Submarine sediment 16, 25
Substitute natural gas (SNG) 40

T

Technical report 64, 66, 74, 80
Temperature swing adsorption (TSA) 137
Triethylene glycol (TEG) 90, 91, 99, 103, 105

U

Urban solid wastes (USW) 43

V

Vacuum pressure swing adsorption (VPSA) 141

W

Water gas shift reactor (WGSR) 46